Routledge Revivals

The Scientific Attitude

First published in 1941 (this edition in 1968), this book explores the relationship between science, culture, and society- focusing on human beings, and human communities. Here, C. H. Waddington uses the concept of science to mean more than factual information about genes and haemoglobin and his subject is the effect of scientific ways of speaking on the ways in which people look at the world around them.

The work discusses biological assumptions made by various communities, particularly fascist movements, on human beings and compares them with the scientific attitude. The Nazis for instance spoke about 'racial purity' and 'German blood' but these expressions, whilst arousing emotion, had, and have, no rational meaning- they are inaccurate and tell us nothing of human genetics.

As well as presenting a scientific argument, being published initially in 1941, this book also acts as a historical document, conveying some of the feeling of living through WWII. It highlights the fact that science and scientific assumptions have very wide implications for the whole conduct of life.

The Scientific Attitude

C. H. Waddington

First published in 1941 by Penguin Books, Ltd.
This edition first published in 1968
by Hutchinson Educational

This edition first published in 2016 by Routledge
2 Park Square, Milton Park, Abingdon, Oxon, OX14 4RN
and by Routledge
711 Third Avenue, New York, NY 10017

Routledge is an imprint of the Taylor & Francis Group, an informa business

Publisher's Note
The publisher has gone to great lengths to ensure the quality of this reprint but points out that some imperfections in the original copies may be apparent.

Disclaimer
The publisher has made every effort to trace copyright holders and welcomes correspondence from those they have been unable to contact.

A Library of Congress record exists under LC control number: 70363607

ISBN 13: 978-1-138-95702-2 (hbk)
ISBN 13: 978-1-315-66538-2 (ebk)
ISBN 13: 978-1-138-95703-9 (pbk)

The Scientific Attitude

C. H. WADDINGTON

Introduced by S. A. Barnett

with a new Foreword by the Author

HUTCHINSON EDUCATIONAL

HUTCHINSON EDUCATIONAL LTD
178–202 Great Portland Street, London W1

London Melbourne Sydney
Auckland Bombay Toronto
Johannesburg New York

First published by
Penguin Books Ltd 1941
This edition 1968

This book has been set in Times, printed in Great Britain on Smooth Wove paper by Anchor Press, and bound by Wm. Brendon, both of Tiptree, Essex

09 088240 7

CONTENTS

ACKNOWLEDGEMENTS

For the illustrations, acknowledgements are gladly made to:

Cahiers d'Art, Paris, 5
Cavendish Laboratory, Cambridge, 10
Cement and Concrete Association, 15
Council of Industrial Design, 12
Salvador Dali, 2
Editions Girsberger, Zürich, 13, 14
Mansell Collection, 11
Museum of Modern Art, New York, 1, 4, 7, 9
John Piper, 6, 8
S.P.A.D.E.M., Paris, 3

INTRODUCTION

In 1941, when this book was first published, the civilisation of Europe was threatened with destruction by the men who then ruled Germany. If the Nazis had won the Second World War, a large part of mankind would have come under a brutal tyranny. Critics of the regime throughout the continent would have been dragged from their homes without warning, imprisoned without trial, tortured and killed. Independent trade unions would have been abolished, and working people would have been at the mercy of the most rapacious of employers. People elsewhere who differed from Germans in colour or physical type would have been treated as inferiors and enslaved. In schools, on the radio, in the press and elsewhere, all teaching and other kinds of information would have been forced into a single pattern. The history taught in European and British schools and universities would have been fiction: Winston Churchill, if mentioned, might have been described as some sort of Communist, in league with wicked Jews to corrupt the purity of the English race.

Movements of this sort were not confined to Germany. Some time before the Nazis came to power in 1933, Italy had come under the rule of the Fascisti, led by Mussolini. These, too, ruthlessly suppressed opposition, and undertook a war of conquest, but less efficiently than the Nazis.

The depravities and absurdities of Nazism and Italian Fascism are not merely past history. The possibility of such a resort to unreason still exists. In Germany today, in 1968, a political party, the NPD (Nationaldemokratische Partei Deutschlands), with a policy which resembles that of the Nazis, has recently won seats in regional elections. In the Republic of South Africa the white ruling minority savagely crushes any attempt of the Africans to achieve even the most elementary political and economic rights.

In this book Waddington contrasts the scientific attitude with the wickedness and irrationality of the fascist movements. Nazis spoke about 'racial purity' and 'German blood'. These expressions, as they used them, aroused emotion, but they had, and have, no rational meaning: they tell us nothing about human genetics or about the fluid that runs in our arteries and veins. But Waddington means by 'science' more than factual information about genes or haemoglobin. *The Scientific Attitude* attracted attention, and provoked discussion, even in war-time when it first appeared, because Waddington refused to be restricted by the barrier which conventionally separates scientists from other people. This is one excellent reason for reprinting the book: the need to widen understanding of what science is about, and to stir up debate on the role of science, is greater than ever before.

Just what is this science or this attitude to which Waddington attaches such importance? Some of the features of the scientific attitude, and of its consequences, can be illustrated by a trivial example. A man grows geraniums in a window box, and finds that they are infested with caterpillars which eat the leaves. He becomes curious to know how the insects came there in the first place, and what makes them eat the younger leaves and flower buds, rather than the coarser parts of the plant. He also wonders what keeps them apart, and how they influence each other.

Some people, when asked these questions, might fall back on replies such as this. It is the *nature* of caterpillars to appear on geraniums; they know what part of the plant to eat *by instinct*; they are spaced out because they have a *territorial drive.* Such 'explanations' tell us nothing useful about caterpillars: they do not enable us to predict when caterpillars will appear, or how they will behave on the plant; nor do they help us to grow more caterpillars, or to get rid of them. One way of describing such expressions is to say that they are not operational: they do not convey information in terms of actual observations. A disadvantage of this is that these statements can never be put to the test of experiment. Perhaps one reason that they are popular is that they can never be proved to be right or wrong.

What does a scientist do? It is sometimes thought that he begins his investigation by collecting a large number of observations as if they were old coins or first editions, and later makes generalisations – theories or laws – about them. But this is not what happens in practice. First, this notion ignores the question: what makes a scientist choose to record one set of facts rather than another? Even if his research is only a hobby, undertaken for recreation, and his material has come before him by chance, the way in which he looks at it depends on the knowledge and prejudices he already has. One observer of caterpillars would be content to record the behaviour of the whole animal, related perhaps to the forms of the leaves, interactions between individual caterpillars, effects of light and dark, wind and rain and so on. Another would prefer to regard the animals as 'mechanisms': he might attempt a study of their sense organs, perhaps in relation to the direction in which light falls on them or odours reach them; or he might attempt a study of their nervous systems; or he might be interested in the mechanics of their movement. All these are possible kinds of investigation, and the way in which any of them is carried out depends on what the investigator has already learnt. Everybody has a framework of beliefs and assumptions, not necessarily conscious, into which he tries to fit each new thing that comes to his attention.

Much research does not even begin like this, but instead is undertaken because somebody 'has an idea'. An example – this time, far from trivial – is the research by Louis Pasteur (1822-95) which finally convinced scientists that living things always come from other organisms of the same sort. It is not long since even learned people believed that mice could be generated from rubbish, and flies from decaying meat. Now, even the smallest micro-organisms are known not to be 'spontaneously generated'. Pasteur undertook his experiments because he was *already convinced* that bacteria (and indeed all organisms) always arise from other organisms of the same kind. If Pasteur had been wrong, his experiments would have ended in a different way, and he would have been forced to abandon his theory. Whenever a scientist makes a guess about how things work, that guess – or hypothesis – has to be tested against the 'irreducible and stubborn facts' of the real world.

The work of Pasteur was not only a major development in science but also of the greatest significance for human welfare. Many important processes, such as the making of cheese, wine, and beer, take place only if certain microbes are present. Still more important, the fertility of soil depends on micro-organisms. And, as everyone knows, many of the most serious diseases can be prevented if the bacteria or viruses that cause them can be kept out.

But Waddington is not especially concerned with these familiar aspects of science. In this book his subject is the effect of scientific ways of speaking on the ways in which people look at the world about them. Both the examples above illustrate two features of the scientific attitude. The first is an intense, often childlike curiosity about the world. The second is that the choice of subject, even by a scientist who is his own master, depends on some initial hypothesis or assumption.

There is also an important *difference* between the two examples. A man whose curiosity is excited by the casual observation of caterpillars is likely to be interested in the diversity of nature for its own sake. He is a 'naturalist'. Pasteur, by contrast, was more interested in establishing valid generalisations: he wished to prove statements about micro-organisms in general, or about the processes of fermentation. He was a 'natural philosopher'. Both methods are necessary, and they may be combined by one scientist.

A natural philospher in this sense does more than observe: he also experiments. This involves interfering, in a systematic way, with the phenomena he is studying. Waddington himself is a natural philosopher; but he does not, as some scientists do, apply his understanding of rigorous methods of experiment only to a few processes observed in the laboratory. He tries to see how this rational, objective way of looking at things can be applied to human affairs. On page 12 he mentions several kinds of society, and he writes: 'Their value is not as models to copy or horrid examples to avoid, but truly as experiments; that is to say as things which give us some insight into the causal mechanisms of social change and thus give us some power to control such changes.' Granted, there are many differences from a laboratory experiment: no individual can regulate what goes on in a human community, as an experimenter might in a population of bees or bacteria; and any scientist who studies human communities is liable to be a participant in what goes on, as well as an observer. Nevertheless, it is possible to decide in outline what one wants, and then to try, quite coolly, to consider what kind of social organisation is most likely to provide it.

In the long run this notion has tremendous implications which have not been generally grasped. Until recently, most human planning was on a trivial scale: a farmer planned his sowing on his own ground; an architect or builder organised the construction of a single building. The activity of a community was usually the sum of the efforts of individuals or of small groups. Today we are accustomed to vastly greater enterprises, but government and industry still make only fitful use of science, except for solving special, technical problems. Yet we could now do much more; this would entail deliberate, conscious planning on the national or international scale, to achieve defined ends.

All this is still about the impersonal, social effects of science. But Waddington writes also about the ways in which science influences the attitudes of scientists and in turn, of others who absorb something of science from their everyday experience. And here we come to the most original feature of this book. Waddington is himself (like many other scientists) interested and excited by literature and the fine arts. He entirely rejects the conventional idea of a conflict between science and the 'humanities'. It is, of course, easier for a scientist to enjoy the arts, than for a non-scientist to understand the sciences.

Unfortunately, many people with a literary education do not merely fail to comprehend the natural sciences: they are positively repelled by them. This is no doubt partly because they have not been equipped, in their early years, to appreciate modern biology or physics. Partly it is the fault of scientists themselves, who often teach even the most exciting parts of their subject in the dullest possible way.

But essential features of the *practice* of science are acute observation and new discovery; and the satisfaction that a research worker derives from an elegant proof or a beautiful microscopical preparation is akin to that of a painter or a poet who has at last achieved the effect for which he has been working. There is a division, but it is not between scientists and artists. The real conflict is between those who create, or at least enjoy, new ways of looking at the world, and those who fear novelty and try to protect themselves from the alarming strangeness of the world by surrounding themselves with barriers. Nearly every technical advance that we take for granted today was resisted, not only by those who stood to lose by them, but also by those who insisted that 'nothing should ever be done for the first time'. Anaesthesia in childbirth, clean water supplies, lighting streets with gas – these and many other innovations were opposed at first; and, since there was no valid reason for opposition, the resisters commonly fell back on some general proposition that introducing them would lead to immorality. Today, modern methods of contraception, humane methods of treating delinquency and much else are arousing similar hostility.

Science represents a constant threat to fixed habits of thinking: scientists, to be effective, have to cultivate a healthy disrespect for authority. A research worker must expect his observations and theories to be critically scrutinised and debated, however eminent he is.

Science is also sometimes equated with inhumanity, and scientists are thought to be cold and lacking in feeling. Everybody who knows a few scientists realises that this is rubbish: and certainly, it is not scientists who resist the use of new knowledge for human welfare; but it is worthwhile to examine how these notions arise. One source is what scientists are actually doing now. A few are perfecting bombs of still greater destructiveness than we have so far seen; others are devising means of killing people with microbes, or of destroying crops and trees, over vast areas, with poison spread from the air. But scientists have also led the opposition to these misuses of knowledge. It would be still better if all scientists, in every country, had refused to take part in these barbarous activities. Scientists, through their special knowledge, have special responsibilities, and do not always live up to them. One reason is the fascination of the technical problems presented even by the design of bombs or the culture of disease germs. (Another is no doubt the need to earn a living.)

But of course the responsibility is not only scientists'. Every atomic physicist and bacteriologist is a member of a community, and reflects opinions and attitudes which prevail in it. As Waddington himself points out, in another context, it is easy to think of an individual or a small group as responsible for some wickedness, but the notion of a whole population involved in ill-doing is harder to grasp, and harder still to act upon. It is right that people should regard modern weapons of war with horror and loathing; but they should not evade their own responsibility by putting all the blame for them on scientists. Nor should they think of the whole of science as destructive, when only fragments of scientific knowledge are used for bad ends.

A more general reason for being repelled by science is that, in his work, a scientist is obliged to try to be impersonal. Most people, shown worms which breed in human intestines, would find them repulsive; but a parasitologist may find them fascinating and even beautiful. Even if he does not, he is obliged (if he studies them at all) to do so without allowing his scientific findings to be influenced by what the worms do to human beings. Of course, his work may eventually lead to preventing infection by the worms; but this possibility, even if strongly hoped for, must not interfere with his appreciation of the facts. The notion that scientists are inhuman comes from confusing this detachment, or objectivity, with lack of feeling for other human beings. A few people with this kind of deficiency can be found in all walks of life, but no one group has any monopoly of them.

More important, scientists themselves have no monopoly of scientific thought or of the scientific attitude. We are accustomed nowadays to talking of scientists as a group apart. This is because, in our highly organised society, science is a profession, and to be a member of it a woman or man must have a set of formal qualifications. But when we examine the behaviour of scientists at work, we find only features which are common, in some degree, to all human beings. I am not referring now only to the desire to benefit other people, or ambition to earn the esteem of colleagues, but also to less obviously social aspects of the scientific attitude. Curiosity and the desire to manipulate and experiment are obvious in all children. Later in life, if these are not crushed by repressive upbringing or by excessively formal schooling, they can become allied to other qualities especially characteristic of scientists who make original contributions to knowledge. Among them are a passion for rational argument and intense pleasure derived from a novel discovery.

Science is not only a narrow kind of knowledge that can vastly improve the economic condition of man; if we wish, it can enter into and enrich every aspect of our lives.

S. A. BARNETT

FOREWORD
by the Author

It is rather a long time since I wrote *The Scientific Attitude;* somewhat over a quarter of a century. It was a very unpretentious little book, which cost only sixpence, when it first appeared as a paper-backed Pelican in 1941. That first version was written hurriedly, because I felt there was something I wanted to say at a time which was undoubtedly one of the great crises in modern British history. It was the period in the early part of the Second World War, just after the British army had withdrawn, heroically but incompetently, from the beaches of Dunkirk, and before either Russia or America had entered the battle against Nazi Germany. It was a time when there seemed a very considerable likelihood that Britain would either be invaded and overrun like France, or would make some sort of sell-out arrangement which would in effect accept the dominant position of the political and ideological system which controlled Germany at that time. It was a very real question in people's minds; what, if any, system of values can we formulate which is convincing enough to sustain the long grim effort that would be necessary to resist the Nazi attack. Had we anything to fall back on, anything of a wider validity than mere nationalist local patriotism?

These are questions which seem rather remote from the world of the mid-sixties. I found it at first rather daunting to be confronted with a request to write a new introduction to the book, rather like being asked to explain and excuse some ancient indiscretion that one had almost forgotten ever having committed. But reading the book again, not only in its first short version, but in the rather longer revised edition that came out in 1948, I find there are some reasons why I do not wish wholly to disclaim it, and am in fact quite glad to see it in print again. For one thing, it is to some extent a historical document. Simply as reporting, it does, I think, convey some of the feeling of living through the war, and particularly the character of the experience that came the way of those scientists who got involved in the running of practical affairs.

It is historical also in another and more important way. It was, so far as I know, the first book in which a practical working scientist stuck his neck out sufficiently far as to assert that science has very wide implications on the whole conduct of life. Science I claimed – and I still do claim – is certainly not merely technological know-how, nor is it only a body of facts and theories about the material universe. It is a method, or manner of dealing with the universe in general, which, if adopted wholeheartedly, can influence the way in which we conduct our entire lives. Science can be the kernel around which there grows up an entire culture, in the sense in which T. S. Eliot used that word when he wrote his *Notes towards the Definition of Culture.* Science has, on this view, a positive contribution to make – and not merely a negative critical one – to such subjects as art, literature, ethics and religion.

In the last quarter of a century, the idea that there is something which can be called a Scientific Culture has become much more widely accepted. But most authors argue that science can, or at least does, provide a culture which is profound only intellectually and rationally, while remaining shallow and ultimately unsatisfying in the sphere of emotion and value judgements. Sometimes this assessment is reached with a certain degree of sadness, as in Snow's writings about the Two Cultures[1] – the scientific and the literary. Sometimes an author who sets off from a basis of science seems to come to the conclusion that he cannot reach the real depth of the meaning of human life without taking off for a flight into the nebulous realm of the mystical and transcendental; Teilhard

[1] C. P. Snow, *Two Cultures*, Cambridge University Press, 1959

de Chardin[1] is an outstanding example, perhaps rather far away from science even in his beginnings. W. H. Thorpe[2], L. C. Birch,[3] Dobzhansky[4] and Hardy[5] – all of them biologists – are writers who start firmly within the bounds of science, but, it seems to me, eventually go so far beyond them that they lose contact.

The point I was arguing in this book is that once you have lost contact with the scientific method – or at least the scientific attitude – you are back in the old miasmic trackless swamps in which man has floundered around for centuries, finding no firm ground to stand on and no steady illumination, but only a bewildering selection of variously coloured will-o'-the-wisps, such as a whole gamut of organised religions, and philosophies including existentialism, logical positivism, Lawrence's thinking-with-the-solar-plexus, and all the rest. I am, of course, certainly not asserting that the scientific knowledge we have today provides a panacea for all man's material and spiritual difficulties. My thesis is that the scientific method demands the participation of all aspects of the human personality; not only the rational, conscious, logical mind, but emotions and unconscious thought-processes as well. In scientific work these various aspects of the personality are mutually critical of each other's operations, so that a real possibility is offered by which an integration of the whole of man's nature might gradually be worked out.

When I made this claim twenty-five years ago I realised, of course, that it was a very bold one – that was a time for boldness; if the Germans had been just a little more successful the book would certainly have been burnt before it was more than a few months old. From the standpoint of today I am even more aware that *The Scientific Attitude* claims rather a lot, and provides only an outline sketch of the arguments and evidence that would be necessary to substantiate the claim. In the years between I have made some attempts to carry the arguments a bit further in a number of articles contributed to

[1] Teilhard de Chardin, *The Phenomenon of Man*, Collins, 1959
[2] W. H. Thorpe, *Biology and the Nature of Man*, Oxford University Press, 1962; and *Science, Man and Morals*, Methuen, 1965
[3] L. Charles Birch, *Nature and God*, S.C.M. Press, 1965
[4] T. Dobzhansky, *Mankind Evolving*, Yale University Press, 1962
[5] A. C. Hardy, *The Living Stream*, Collins 1965; and *The Divine Flame*, Collins, 1966

various journals or symposia.[1] I have also developed the theme of chapters III and IV, which deal with the relations between science and art, into a full-length book which is shortly to be published.[2]

The main argument I have had to try to reinforce is that which still asserts that science can assist in providing a basis for ethical judgement. It is one of the most sacred dogmas of philosophy as taught in the schools and universities that facts and values belong to different realms and have no essential connection with one another. Some of the reviewers of *The Scientific Attitude* bluntly stated that when I asserted that science (which deals with facts) can have something to say about values, I was committing a blunder in elementary logic. I think the logic they were talking about is so elementary that it is best forgotten. Logic is the science of the necessary connections between ideas, and cannot assert anything about the real world, which is inhabited not by ideas but by existing entities. To state that there is no logical connection between facts and values is merely to tell us something about the way in which you define the use of the word 'fact' and the use of the word 'value'. If you define a fact as not including anything to do with value, then of course there is no logical connection between these two categories; but there is no reason why you should define facts in this way – except perhaps some historical hangover from the notions used in the early history of the science of physics. It took me some time to get this point clear in my own mind, and to get rid of the pervasive influence of the simple materialist ideas about nature which were the basis of Newtonian physics. I first argued the matter out in public with a number of critics, who included philosophers, psychologists, bishops as well as other scientists, in a whole series of letters and replies that were published soon after the first edition of *The Scientific Attitude* as a book

[1] *The Ethical Animal*, Allen & Unwin, 1960; paperback, Chicago University Press, 1967. *The Nature of Life*, Allen & Unwin, 1961; paperback Unwin Books, 1963. *Behind Appearance* (a study of painting and science in this century), Edinburgh University Press and M.I.T. Press, to be published shortly. 'Mobilising the World's Biologists (towards the International Biological Programme)', *New Scientist*, 1963. 'The Desire for Material Progress as a World Ordering System, in Conditions of World Order', *Daedalus*, Spring 1966

[2] *Behind Appearance*, Edinburgh University Press and M.I.T. Press, to be published shortly

entitled *Science and Ethics*.[1] I returned to the theme in a much more thorough and, I think, clearer way in my book *The Ethical Animal*.[2] Perhaps I can best indicate the directions that my thoughts have taken about this matter of the relation between facts and values by quoting the first paragraph in the preface to that book.

This book is an attempt to establish a certain thesis about the nature of the framework within which our ethical beliefs should be evaluated and criticised. To summarise the argument very briefly, I shall try to support the following four points: firstly, that the human system of social communication functions as such an efficient means of transmitting information from one generation to the next that it has become the mechanism on which human evolution mainly depends. Secondly, that this system of 'socio-genetic' transmission can operate only because the psychological development of man is such that the new-born baby becomes moulded into a creature which is ready to accept transmitted information; and I shall suggest, it is an empirically observed fact that this acceptance is founded on the formation of 'authority-bearing' systems within the mind which also result in the human individual becoming a creature which goes in for having beliefs of the particular tone that we call ethical. Thirdly, I argue that observation of the world of living things reveals a general evolutionary direction, which has a philosophical status similar to that of healthy growth, in that both are manifestations of the immanent properties of the objective world. Finally, I conclude that any particular set of ethical beliefs, which some individual man may advance, can be meaningfully judged according to their efficacy in furthering this general evolutionary direction.

This is an aspect of *The Scientific Attitude* in which, as this quotation shows, my wish today would be to proceed further along the same direction as I had begun to sketch out when I wrote that book. There are other aspects of it in which my present opinions are more at variance with the ideas I put forward twenty-five years ago. Some of these are matters of relative detail or particular phraseology. There are, however, four major topics on which I should like to say something to the reader before he plunges into what I wrote quarter of a century ago. These are the attitude to fascism and communism, and to religion, the question of underdeveloped countries with the associated problem of racism, and finally the question of world peace.

[1] *Science and Ethics*, ed. C. H. Waddington, Allen & Unwin, 1942
[2] *The Ethical Animal*, Allen & Unwin, 1960; Chicago University Press, 1967, (paperback)

The discussions of Fascism in chapter VI and of Communism in chapter VII will both I daresay strike the young reader of today as rather quaint and old-worldly. Well, all right; remember that reading this book should be taken at least partly as an experience of past history, a contact with what people much like yourselves used to think some years before you yourself were born onto the scene. The flavour of the old-fashionedness which I suspect the modern reader of detecting will probably differ in the two chapters. At the time they were written we were actually engaged in a war against Fascism. The prevailing fashion among the more hard-headed individuals was to conceive of Nazism mainly in terms of its practical actual operations and factual results in the world – the provision of more or less full employment, the production of a war machine which cut through Poland and France like a hot knife through butter, the burning of the books, the obliteration of modern painting and architecture, the driving out, followed by the actual genocide, of the Jews. I happen to have worked in Germany as a research scientist in the years just before Hitler came to power. I therefore knew something about the cultural background out of which Nazism sprang and which it so quickly replaced. The songs of Bertold Brecht and his *Drei Groschen Oper*, novels like *Berlin Alexanderplatz* by Alfred Döblin, *Regierung* by B. Traven and *Fabian* by Erich Kästner, were part of the mythology of my youth (does anyone read them any more?) and provided a certain counterpiece to envisaging the Nazi system as simply the operations of an authoritative economic and military planning staff. In the years since the war it is the Nazi *mystique* which has attracted more attention, for instance in such a work as Hannah Ahrendt's book about the Eichmann trial,[1] and – at of course quite a long remove from the political scene – the movement of enthusiasm for what I should like to call a 'refined blood-and-guts outlook' of the D. H. Lawrence-Leavis vintage. However, I should still claim that, although *The Scientific Attitude*'s line that the *mystique* is bunk and all that matters are the factual consequences, may seem a bit out of date and is certainly not wholly adequate, it cannot be entirely dismissed either: there is something – even quite a good deal – in it.

[1] Hannah Ahrendt, *Eichmann in Jerusalem*, Faber & Faber, 1963

The discussion of Communism must, I think, be judged nowadays as being slanted in exactly the opposite direction. In the thirties and right up to 1940 when this book was written we tended to pay too much attention to the *mystique* of Soviet Communism and not enough to its actual results. Of course, at this time the Stalinist purges had not yet reached their height and in Britain we knew very little about them—much less than we knew about the Nazi atrocities. Again, in extenuation one may point out that the material results of Soviet Communism were by no means to be despised. A vast country inhabited by an enormous population, most of whom had been downtrodden serfs and debased industrial workers, with a minute surface froth of middle-class intellectuals, and a still more tiny scum of extravagantly rich aristocrats, had been converted into some sort of simulacrum of a modern industrial society. Overall, this amounted to a very considerable increase in the well-being of the human species. We may well ask nowadays whether a similar increase in well-being would not have been achieved by any other system which brought about the industrialisation of Russia, independently of any Marxist ideology – and possibly done so even at a less staggering cost in human suffering (though nobody, not even the Japanese, have actually discovered how to get themselves up to and through the 'take-off point' into the modern world without suffering a lot of agony in the process). But the intellectuals of the Western world during the thirties lived in a system within which there was both an obvious gross failure to mobilise the resources of society for the good of its members – this was the time of mass unemployment and the Hunger Marchers – and also an explicit denial by the established authorities that scientists or other intellectuals had anything useful to say. The Soviet Union may not have shown much more economic competence than we did, considering the vast unexploited natural resources which it had to tap, but at least it did officially say that it approved of science and felt that scientists could be of help. We were, I suppose, suckers in swallowing more of this line than we should have, but again the fact that we did so is part of history.

Now about religion. *The Scientific Attitude,* when it mentions it at all, dismisses it airily as something which no sensible person is really interested in. This is very far indeed from the climate of opinion among young people nowadays. I am bound to admit that I do not find it possible to find in myself, even today, the possibility of going very far with the modern revival of religious feeling, but there are some remarks I should like to make.

In the first place when I mentioned religion in *The Scientific Attitude* I was thinking of officially organised churches. Moreover I was writing in a period before there had been any noticeable impact of the modern types of theology – before the meaning of religion had been remodelled by people like Martin Buber, Richard and Reinhold Niebuhr, Karl Barth, Paul Tillich, let alone Mervyn Stockwood or Teilhard de Chardin. If religion has become more concurrent with the thought of young people nowadays, this is as much or more due to changes in religion than to alterations in what young people think now as compared with what their fathers thought twenty-five years ago.

I am, however, prepared to contemplate the possibility that there may be more to it than this, and that I may genuinely suffer from a blind spot about certain aspects of religion – those that see the religious experience as some sort of relation between the human being and a *personal* Super-Entity. Perhaps I may be excused a word of autobiography. I am the descendant of a long, considerably inter-married network of lower middle-class Quakers – Waddingtons, Warners, Cappers, Abbotts. As a pre-adolescent child my religious experience consisted of two things. At a relatively minor and not very impressive level, there was a rather rigidly controlled Sunday – no card games or draughts with my sister, no lead soldiers, only approved books to read, a formal lunch with a roast joint on a clean linen table-cloth, and if you put your elbows on the table somebody would pick your arm up and bump the funny-bone on the table to remind you to mind your manners, and after lunch my grandmother, head of the household, read out a chapter of the Bible. Of deeper effect was the Sunday morning experience of attending a Friends' Meeting. We walked – rain or shine – a mile and a half to a little unpretentious hall in a back street of the small market town of Evesham in Worcestershire – passing on the way, I remember, a place called Battledore Bridge; the river Avon had run red with blood down to this point when some beastly English king suppressed the democratic rebellion of Simon de Montfort at a battle way up the river at the other side of the town in 1400 or 1500 and something. At any rate that's the mythology I carried in my eight-year-old head. The meeting itself took place in a more or less square room with three ranks of benches around all four sides of it. When the Elders of the Meeting decided it was 11 o'clock, we all went in and sat down, and there we sat until they decided it was 12 o'clock, when they turned to one another and shook hands, and we could all get up. For the most part we sat in dead silence, but at any time anyone who decided he had something to say which others might like to hear he was at liberty to get up and say it. Probably on most Sundays there would be two or three people who would talk, usually for not more than about ten minutes each. They were quite intimate talks, of people all of whom had been going to this same Meeting House for ten, twenty or more years, many of them with families connected by marriage. Now the whole slant of the Quaker *ethos* – at that time at least, I am not sure if it is still the case – was almost completely dissociated from belief in any specific dogmas. The emphasis was entirely on the contact between the individual human being and the general structure of the Universe – a structure, of course, not at all merely matter of fact, but embodying mystery and value – 'numinous' to use a favourite word of my old friend Joseph Needham – the world inhabited by Wordsworth and Thoreau, to mention an author who was amongst the approved writers whose books I read at this time. But the point in the present context is that there was, in this Quaker experience in a small town in the depths of the English countryside, no trace either of a formulated system of theological beliefs laid down by ecclesiastical authority, nor even very much suggestion that the deeper realities of existence with which one was expected to make contact during the long periods of silence, sitting in company on the benches around the walls of the Meeting House, was in any way *personal*, to be conceived of in terms of a Father, a young man who suffered, or a virgin Mother.

It is from this background that there emerge any comments

I may make in *The Scientific Attitude* about religion. I should, I suppose, always have qualified it as 'organised religion'. Since what I was thinking about were things like the Thirty-nine Articles or the Dogmas of the Catholic Church. Such formalised systems of tenets are, I am still convinced, almost always an impoverishment of the direct contact with the ultimate mystery of existence which the Quakers of my youth were trying to attain. I am willing to contemplate the possibility that they may perhaps in their turn, have impoverished their conception to some extent by failing to attribute personality to it in any meaningful sense. However, I am very conscious of the danger of sliding away from the very difficult endeavour to retain a direct contact with the ultimate mystery of existence – best done in silence without verbalisation, as in the Quaker Meeting – into any form of explicit dogma, or emotional relation with a Super-Person. Is it 90%, or only 75%, of the crimes carried out by man against man in the last few centuries which have been justified in terms of 'Religion'?

The poem by Paul Eluard which I quote on page 45 was written about a person, a girl, with whom he was obviously expecting to go to bed (again), probably without benefit of clergy. Does this make it any less 'religious' than 'Onward Christian soldiers', or such casebook Freudian items as 'Rock of ages, cleft for me'? (Perhaps I should have pointed out that, after the somewhat idyllic religious initiation of the Friends' Meeting House in Evesham, I went to a boarding school and did my stint of clocking-in at Chapel at 8.50 every morning, with roll-call, before getting into the classroom at 9.10. Believe me I know what Religion means – if one has to show sympathy with the aspirations of the young people of today – at least what it meant when this book was written.)

Then we come to the three deeply interwoven and interacting themes of the developing countries, race feelings and world peace. These issues are, rightly, in the very centre of interest of people today. If *The Scientific Attitude* had been written yesterday, the nonchalance with which it skates over them would be inexcusable. But, remember, it is a historical document, written in the pre-history of the world of the sixties – in its original version, before even India had become independent or any atom bomb had exploded over Nagasaki,

I did not say very much about these issues, but believe it or not, it was more than most contemporary writers were saying at that time.

Rather surprisingly it is on the most emotionally laden issue in this whole tangle of subjects that science can speak with the clearest voice. Race has come to be a dominating political and moral issue in the world today, chiefly because of a reaction against what was – and still too often is – a widespread tendency to form a pre-judged opinion about the value of any individual on the basis of his race or colour. Now science can prove quite unequivocally that none of the external appearances which give an indication about a person's racial origin – the colour of his skin, the character of his hair, the form of his facial structures, and so on – give any indication at all of his intellectual abilities, emotional depth and maturity, or his general character structure. There are clever and good people with black, brown, yellow and white skins, just as there are wicked or evil people. A first glance at a person which 'places' him racially – or even a more prolonged examination, to determine his racial affinities – does not tell us anything about the more fundamental characteristics which determine his individual worth. These can only be assessed by looking precisely for the qualities which we consider important in human life.

The Scientific Attitude does not, as I have argued throughout this book, deny the reality of the concept of value. I would not deny that some people are better than others. Obviously there are many different characteristics which contribute to the value of a human individual: intelligence, knowledge, enterprise, reliability, emotional depth and stability are among them, and there are many others. No one individual is likely to score very highly on all the possible desirable attributes of human personality. The first point to be made in connection with race is that if you compare two groups defined as belonging to different races, such as white Englishmen with coloured West Indians, estimating the numbers of individuals which score at different levels in tests for one of the desirable human characteristics, such as intelligence or emotional maturity, then in all cases that have ever been studied it is found that the two groups overlap to a very large extent. This is merely another way of stating the same fact that was made in the previous paragraph.

However, there is another question to be raised. It may be that the average score for some desirable quality is higher in one group than in the other. There is little doubt for instance that the average score of a group of Englishmen would be higher than that of a group of Africans from the Congo on such subjects as a knowledge of the basic rules of hygiene, and probably in such valuable qualities as enterprise in improving their conditions of life, and readiness to co-operate with one another in carrying out the necessary tasks. There can be little doubt that, *as they exist at the present,* some groups are in a real sense better on the average than other groups at some things, even though one cannot, as we have just seen, deduce from this difference in averages anything about the relative merits of particular individuals.

But one has to go on immediately from this statement to make two other points. The first is this : if one compares two groups on their average abilities in a number of different characteristics it will scarcely ever – in fact probably never – be the case that one group is consistently better all along the line. It may have an advantage in intellectual ability, knowledge, tenacity of purpose and so on. But the others may score in a different range of qualities, such as kindliness, generosity, artistic ability and so on. Anthropologists who study the more backward peoples of the present-day world always seem to find that even the most apparently uninspiring peoples actually have many good and praiseworthy characteristics. A more profound question, though, is to enquire, not what peoples and groups are like at the present day, but what they could become given a different system of education, social organisation and material amenities. This is the crucial question in much racist theorising, which has attempted to argue that some races are inherently and biologically inferior in their potentialities to others.

The scientific approach to this question operates at two levels. At the first, it seems to be rather equivocal, and many people find it emotionally unsatisfying. At this level the scientist has to say that we know that in other widespread species of animals there are local populations which differ in the variety of hereditary potentialities contained in them. In fact this situation obviously occurs in man also, and is very obvious in relation to certain superficial physical characteristics such as skin colour; no individual among a group of natives of Scandinavia is likely to contain the hereditary potentiality to develop the skin colour of a Central African. But the important question is, do we find such differences in the distribution of human potentialities required for the development of really valuable human characteristics, such as those mentioned above. Are there human groups whose hereditary potentialities for intellectual argument or mathematics or music, etc., etc., are on the average lower than that of some other group? It has to be admitted that biologically this is quite a conceivable possibility. I think the overwhelming majority of students of human heredity would at the present time have to admit that they really do not know whether such differences in average hereditary potentiality do or do not occur in the world as it is at present.

This may seem a very unsatisfactory answer, and it may seem to leave the door open to prejudiced people who wish to argue that some group, such as the Negros, are inherently inferior in hereditary potentiality for some valuable types of human attribute. However, the science of human genetics can proceed to make a second point which almost entirely removes the sting from its previous confession of ignorance. Why, in fact, is it that science has not yet been able to determine whether these are differences in important hereditary potentialities in different groups of mankind? The reason it has not been able to do so is that there is no feasible method of testing the full extent of human potentialities for the kinds of qualities which we find most important in human life. Even potentialities for simple physical qualities such as growth in size, height and weight are difficult enough to determine, since the amount of growth actually realised depends enormously not only on the hereditary potential but on the nutrition and conditions of life while the individual is growing up. There has been a steady increase in the average height of people, even in such flourishing countries as England, over the last century, while in countries which have had poor nutrition in the past, such as Japan, the change has been quite spectacular. When we consider the more subtle qualities of personality, such as knowledge, intelligence and all the other aspects of psychological and spiritual life, experience shows that in most of them the influence of the environment is even more powerful. There may be some mental abilities which are rather directly controlled by heredity and on which pupation and training have little effect; possibly great musical ability, or alternatively tone-deafness, may be of this kind, and probably the fantastic abilities of some exceptional individuals as 'calculating boys' are examples. But in general, for most of the qualities we consider important in a person's character, education, training and circumstances are of almost overwhelming importance.

The most important thing that follows from this is that we do not yet know the full hereditary potentialities of even the most highly educated and fortunate groups of the human race in existence today. There is no reason at all to believe that even the most educated groups and the most flourishing industrial nations have reached the ceiling of what their potentialities are capable of. Even less, of course, can we suppose that people who have grown up in less favourable circumstances are expressing anything approaching their complete potentialities. It becomes senseless to ask whether group A is potentially inferior to group B, when we know that we are only seeing a small fraction of the potentialities of either of them. We can be pretty confident that in the world today every human group (which has above a fairly modest number of individuals, say 10,000 upwards) contains hereditary potentialities which could be exploited to bring the average performance of that group above that of the present-day average of the most favoured countries existing now. There is no human group – except again the very small ones – for whom their genetic potentialities need be a factor limiting their advance in the foreseeable future.

This is a situation – paradoxical enough to those who do not have The Scientific Attitude – in which our knowledge of how ignorant we are removes a fear that might stand in the way of the advance of mankind.

The problems of the developing countries, which have come to be the centre of interest for most forward-looking people at the present time, are in general so interwoven with political issues that I would not have discussed them at length, even were I writing *The Scientific Attitude* for the first time today. I think, however, one must say that the scientific approach would, in the first place, stick firmly to the banal, matter of fact, point of view that what really matters in the developing countries are such down to earth facts as the mean expectation of length of life at the time of birth, or the excess of the gross national product over the level of their subsistence. The primary essence of 'development' of a country is that a woman – and let's start by agreeing the elementary point that a woman's feelings are as important as a man's – should be certain that she can rear to maturity almost all the babies she bears. A socio-economic system in which something like half the babies born will die before they are aged five, as is the case in many underdeveloped countries today – imposes on its members a degree of emotional superficiality which almost rules out of court anything which we and the fortunate Western countries would recognise as a modern civilised personality. Of course, you can have wonderful civilisations based on personalities with these blind spots of insensitivity. Classical Greece was almost totally insensitive to a substructure of slaves, to the total subjection of all women except a few fascinating courtesans, and to a high level of infant mortality which they did not mind aggravating with a bit of infanticide when it seemed desirable. In Renaissance Italy, or Elizabethan England, the fact that a few tough guys like Benvenuto Cellini or Sir Walter Raleigh got through to the top was considered enough to justify all the miseries of the lower classes which supported them. It is, I feel sure, a real advance that at the present time we are not willing to accept arguments of this kind. In fact it is *the* real advance. What we have to insist on, in my opinion, is that in every region of the globe the social and political system should make it as easy as possible for everyone – male and female, clever and stupid, tough or weak – to develop as far as possible every potentiality that they have in them.

Edinburgh, 1968 C. H. WADDINGTON

SCIENCE AND CULTURE

the terrific labour and unpleasantness involved in many of the most important Nazi ideas, such as its desire for world dominion. But even this idea, with all the toil and suffering that it brings with it, appeals to something in many people – certainly in many Germans; and it was presented to them along with other ideals which probably appealed more strongly and, at the beginning, seemed more immediately practical and therefore more important.

A new system of social ideas, such as those of Nazism, does not have much chance to be taken up and put into practice unless people are suffering more than usually from the stresses and strains of life. In Germany before Hitler came to power many people were certainly unsettled enough to be ready for a change of some sort. Their difficulties were of two kinds, economic and what one might call general. Now of these two the economic may, very likely, be the more fundamental, and the intellectual cynicism and disintegration of the whole system of values and morals, which was so noticeable in Germany, may in the last analysis be a result of the misfiring of the economic machine. But for most people, things don't appear quite as the economists suggest. A man's economic difficulties are straightforward affairs, such as being out of work, or trying to cadge a living as a car salesman although he has an engineering degree. The connection between these circumstances and his moral and emotional troubles will probably be clear enough; his family life may break down because he can't afford to give his wife a decent home, or his son may take to drink, or more recondite vices, for sheer lack of anything else to do. But what is in theory important, and what in practice does not enter into one's ordinary experience, is the connection between social disintegration and the particular kind of breakdown which the economic machine is undergoing. It may, or again it may not, be true that you cannot have a system of lifelong monogamous marriages unless the economic system provides for at least a certain amount of private property. But even if it were absolutely true under all possible circumstances, nobody would know it as an immediate fact of experience. And if people are starting to recast the social system, it is likely that they will first decide that they want certain social and emotional things, such as monogamy or freedom of opinion, and only come later to the conclusion that they must have an economic order organised in corporations or limited liability companies or whatever it may be as a means of attaining their primary non-economic aims.

It is very important to remember that people may easily be mistaken as to the appropriate economic organisation for attaining their general aims. Such a mistake was made by the German people. The reason why they were willing to support Nazism originally was that they wanted certain things in their daily lives; some of these things, such as a system of established values which they could accept without question, and a sense of belonging to a community, they probably got; but others, such as a better standard of life, and security, they did not.

It is for these reasons – because non-economic considerations are a motive force in human behaviour in a more immediate and conscious way than different systems of economic organisation are – that it seems justifiable to discuss the modern world with an emphasis on its cultural aspect rather than its economic. By doing so I do not imply that the cultural motives were the most important for political leaders, as opposed to their followers. And it will be apparent all through the discussion that the two aspects cannot be at all sharply separated, and we shall come back later to the question of what the cultural requirements demand of an economic system that will allow them to be fulfilled.

The dominant characteristic of the cultural life of the world today is the existence of revolutionary movements. Communism is one, already over a quarter of a century old in the country of its origin, but still with its old fire in its newly conquered territories, and something not very different even in Russia herself. Nazism was another movement whose adherents, at any rate at the beginning, felt they were playing a part in a movement which would change the course of history. In England and America we know little of the vivid self-confidence and impatience of second thoughts which the revolutionary ardour produces; and little also of the difficulty for a single individual living in a community filled with this spirit to have any ideas of his own, independent of those surrounding him. But in large areas of the world – perhaps the greater part of it – men are living in consciously revolutionary societies; and even in the more sophisticated and self-critical Western European civilisation, people have had a taste during the war of feeling themselves a part of an all-embracing social effort, and alongside the relaxation of going back to ordinary life there is a feeling that things have grown smaller and of less consequence. Moreover we have seen a bit too much of the demerits of all the various revolutions offered to us – Nazism, Communism, 150% Nationalism – to feel much confidence in any of them. But a revolutionary set of ideas cannot, at least after it has attained a certain impetus, be defeated by a mere blanket of inertia. If Western Europe is to retain its old cultural importance in the new world of the future it must discover a new outlook to guide and animate its life.

The ideas on which our industrial civilisations have been run are breaking down everywhere, not only in Germany. They went first, and most completely, in Russia, in which a capitalist industrial system of the English type was in fact never fully developed. Then they were superseded in Italy, where again they had never been fully developed. The great slump of the beginning of the 1930's not only swept them out of Germany, but loosened their hold in England, France and America, where they were most firmly fixed. Who believes now that we shall ever return to the economic system and the social ideas of 1913? That world has gone; and we cannot oppose the chaos and anarchy of today by calling for a return of it. And yet, it is difficult to deny that England now is a worse country to live in than it was then.

It is not only that before the war we were beginning to accept mass unemployment as nothing out of the ordinary, and were getting accustomed to the fact that in times of peace our productive resources are sometimes less than half used. But, with the acceptance of such a low standard of social efficiency, we were also tending to acquiesce in a gradual loss of social enthusiasm. Before the war most people were half persuaded that the old pride in being an Englishman was after all mere imperialistic jingoism, an unworthy attitude in face of the international problems of the day. That was perhaps in a sense true. But no society is worth much if its members cannot find something which makes them proud to belong to it. It is one of the conditions for the success of a new social order that it supplies something which fills this need, and it is one of the conditions of our survival that we shall find something which will make us as proud to be democrats as Nazis were to be Nazis.

There will never be any great enthusiasm for belonging to a society which is not going forward. And throughout England in the years before the war progress was at a discount. In almost all walks of life the way to get on, to reach positions of power and responsibility, has not necessarily been by the exercise of initiative and drive, or by showing an ability to seize opportunities and make the most of them. It has been at least equally useful to demonstrate that one knows the correct thing to do.

Our culture is full of unformulated rules of conventional behaviour, and we have placed a perfectly absurd value on the ability to conform to those rules, and thus to preserve the whole system whose behaviour they are supposed to regulate. A few years ago, many people said that one of England's greatest leaders in the last twenty years was the former Prime Minister Mr Baldwin, the late Lord Baldwin of Bewdley. And they said this, not so much because they believed that he created any great new values for us, or saved us from any major disaster, but because of the way in which he handled the abdication of Edward VIII. That is, what impressed them was Baldwin's knowledge of what would seem polite behaviour in an embarrassing situation; just what note to strike over the wireless, just how to bring the Dominions into the situation, and what gestures should be made by the Church and Parliament and House of Lords and so on. All these things are actually perfectly trivial compared with facts of economic organisation and international politics. Baldwin is also the man who, running on a 'Trust Baldwin' platform, consciously misled the people as to his views on the international situation in order to reach a position in which he could carry out a rearmament which he believed to be necessary; and, having reached the position, neglected to carry out the rearmament to a sufficient extent. Only a society which is completely sunk in formalistic politeness could consider that of less importance than the ability to handle the ritualistic niceties of exchanging one king for another. England in recent years has been very dangerously near to this. In little things, as in big ones, the man to whom responsibility and trust was given was only too often the man of experience in keeping things as they are and not the man of initiative in improving them. During the war we did quite largely succeed in getting away from this attitude; and the election of a Labour Government showed that the country as a whole was prepared to put ability and initiative before correctness. But Britain is still, thank Heaven, a country of tradition; and it is as well to recognise the dangers, as well as the advantages, of that.

There is no denying, of course, that tradition has a value, and a very high one. A great many people who felt like running into a hole as soon as the air-raid sirens started up were kept going mainly because they knew it wouldn't be the thing. And the refusal of England to panic at the time France fell was mainly due to our stubborn acceptance of the traditional idea that England doesn't get beaten. I doubt if we should become hysterical, as many Americans did some years ago, at a realistic broadcast of an invasion of the world from Mars; such things just don't happen here. The strong belief which we still have that there are certain ways in which things should be done, and others in which they should not, is one of our strengths. But it becomes a weakness if the ways we believe in are too petty and rigidly fixed, and to some extent that is the case. We emphasised the importance of 'correct' and traditional behaviour to such an extent that we almost lost the art of dealing with new situations which tradition did not foresee. Sometimes we get away with it by a sudden burst of initiative when things get really bad – our famous 'muddling through'; it may come off, but again, as in the early part of the war, it may most definitely not.

It took several years of war before we learnt to sort out the really competent men, give them a reasonably free hand, and thus produce what the American Supreme Commander admitted to be the most efficient planning staff on our side.

The low estimation in which initiative is still held in England is probably hardly realised by most people who have never experienced anything else. But the contrast is very striking in America, for instance. People there are much nearer to a time at which initiative was a real social necessity and brought its own reward. When America was pushing out to the west and opening up new country, the drive and work of an individual who brought new land into cultivation or who founded a new business brought new wealth into the whole community; and the social values appropriate to those conditions are still recognised to a considerable extent in America, although the conditions of social life now are actually very different. As a matter of fact, a similar process of development must have gone on in England in a less obvious way. Our economic system was, in the early years of the last century, based on individual competition between small businesses, and originality and efficiency were qualities which an individual could show with his own resources; and they brought success, and were prized for doing so. Nowadays even in America there are hardly any empty niches left in which an energetic and competent man, starting from scratch, can raise himself to the top of the social pyramid; in spite of the ballyhoo about the 'Land of Opportunity', the great majority of American labourers will remain labourers to the end of their days. And in England the independent producer has almost ceased to exist. The work of society is carried on, not by self-sufficient individuals, but by large firms and corporations. It is becoming less and less possible to set up on one's own except as a professional man or a shopkeeper, and even then the conditions of work are very largely controlled by external authorities. Most people today have to enter a firm in a junior position as wage or salary earners; and getting on is not so much a matter of doing something real, with things, as of doing something psychological, to impress one's bosses.

It is probably not impossible to find a social system in which the integration of production into large units does not curb and frustrate the initiative of the men and women concerned. Certainly one must hope that it is possible, since there is no doubt that the integration is here to stay. In fact, it is clearly increasing all the time, and will increase further. The system of free enterprise of small units, whatever its merits were in the past, has been failing to work for many years. It failed even as a money-making concern, and we have seen a fairly rapid absorption of the small businesses into enormous combines and monopolies. The main driving force of these has been, of course, to make profits and distribute dividends; and they have mostly found that the best way to do so was to restrict output and charge high prices. But the need for centralised or monopolistic control has been at least equally clear when the purpose in view was production, as it may be in war, and not merely profit. In the last war the greater part of the economic system of the Allies was immediately put under fairly close government control, and as the war progressed we found it necessary to go even further in centralisation. It is clear that whatever happens we cannot return to unorganised small-scale businesses unless we are willing to put up with a very much smaller volume of production and a very much lower standard of life than would be possible to us.

The movement towards large-scale organisation of productive and business enterprise is so obvious, and has been so often pointed out, that it is hardly necessary to bring forward evidence of it. But it may be worth while quoting, as an example of a technical but non-political opinion, from the report of the committee set up by President Roosevelt in 1937 to report on the effects of technological changes on national affairs.[1] 'A return to small-scale production methods in industry and agriculture cannot help but curtail technological progress at this period of history. Scientific research upon which modern technological invention is based has too many ramifications and is too costly to be undertaken and financed by small producers. Inevitably, existing technology, which is the primary contribution of the Western world to civilisation, could not be maintained; its continuance has already met strenuous opposition wherever the theory or practice of small-scale production is current.'

If business is organised on a large scale, many of the conditions of life will be determined by agencies whose powers extend over wide areas of the earth's surface. This does not necessarily imply that all governmental control must be even more centralised than it is now, and exerted more uniformly over the area which it covers. It is probably quite possible to centralise the administration of only the major aspects of policy, and to leave to smaller geographical regions, such as Scotland or Wales, more control over details than they have at present. Tendencies towards this kind of organisation are already apparent even in the highly centralised states such as

[1] Superior figures refer to end-of-text references (pp. 147–9).

wartime Germany and Russia. In the former there was a certain independence of the 'Gaus' into which the country was divided; while in Russia the constituent republics of the Union have a considerable autonomy in cultural affairs.

Even if some balance can be struck between centralisation and local autonomy, man cannot expect to return to the isolation and independence he used to have. In the old days a man could consider that his opinions and his intimate habits of life were nobody's business but his own. At the present day some one or other is continually getting at him to change them; either the government lays down rules as to what he should think, or newspapers, cinemas, radio, posters bombard him with suggestions.

This continual intrusion into what used to be considered the sacred privacy of the individual is again primarily a technological and only secondarily a political change. It can only be repelled, and the trend reversed, at the cost of discarding our modern methods of production. If we want to keep our lives inviolate, we shall have to go back to a civilisation in which we only meet our neighbours occasionally, after a long plod through muddy lanes to market.

These two results of technical development – large-scale organisation, and intrusion into the privacy of the individual – are, when they are given a political form, the hall-marks of what are usually called totalitarian systems. The meaning of this term is also coloured by our feelings about the countries in which such political régimes have been set up – hatred for the Nazi and Fascist systems, and usually at least scepticism about the Russian. But the objectionable features of these states are not inevitable accompaniments of totalitarian economic systems, as one can see from the fact that the three differ among themselves. The totalitarianism, on the other hand, does seem to be inevitable; the whole trend of recent history is towards it. One cannot dismiss the Nazis, Fascists and Communists all together for being, in their different ways, totalitarian; fairly soon we shall all be so, in some way or other. The totalitarians of today have taken, with the wrong foot foremost, a step which we shall all have to take tomorrow.

Instead of harping on their similarities, we should consider very carefully their differences. From a purely economic point of view, they present three full-sized experiments in possible methods of organising the productive forces of a country. If we have got to do that ourselves, as it seems obvious that we have, their preliminary efforts must provide a vast number of data from which most valuable guidance can be drawn. But this is a task for technically competent economists. From the point of view of the ordinary man, the Fascist-Nazi régimes, and the Communist régime, are two experiments in building social systems founded on organised totalitarian economics but with explicitly stated social aims which are completely different. Again there must be very valuable evidence about questions which one does not need to be an expert to understand. Is it possible, for instance, to combine totalitarianism with freedom of thought, or with initiative in execution? How can the most able men be picked out for authority?

Before we examine these systems to get what help we can from them, we must know what we want from them. It would be very unenterprising merely to look them over and select any features which happened to appeal to us. Their value is not as models to copy or horrid examples to avoid, but truly as experiments; that is to say, as things which give us some insight into the causal mechanisms of social change and thus give us some power to control such changes and cause them to proceed in the way we desire. That is the essential nature of a scientific experiment. When Newton first passed light through a prism and produced a spectrum, he made a very pretty picture, all the colours of the rainbow. But the significance of the experiment lay not in the production of a pretty effect which other people could repeat, but in revealing something new about the nature of white light, namely that it is made up of a mixture of light of many different colours; and this is important because it increased one's power of using light for other purposes which one may have in mind.

men, the basic characteristics which society cannot change but must accept and build on, are much vaguer than anyone could have expected.

'Differences between one animal and another, between one individual and another, differences in fierceness or tenderness, in bravery or in cunning, in richness or imagination or plodding dullness of wit – these provided hints out of which the ideas of rank and caste, of special priesthoods, of the artist and the oracle, could be developed. Working with clues as universal and as simple as these, man made for himself a fabric of culture within which each human life was dignified by form and meaning. . . . Each people makes this fabric differently, selects some clues and ignores others, emphasises a different sector of the whole arc of human potentialities. Where one culture uses as a main thread the vulnerable ego, quick to take insult or perish of shame, another selects uncompromising bravery. . . . Societies such as those of the Masai and the Zulus make a grading of all individuals by age a basic point of organisation. . . . The aborigines of Siberia dignified the nervously unstable individual into the shaman, whose utterances were believed to be supernaturally inspired and were a law to his more nervously stable fellow-tribesmen. . . . A people may also, like the BaThonga of South Africa, honour neither old people nor children; or, like the Plains Indians, dignify the little child and the grandfather; or, again, as with the Manus and in parts of modern America, regard children as the most important group in society. . . . No culture has failed to seize upon the conspicuous facts of age and sex in some way, whether it be the convention of one Philippine tribe that no man can keep a secret, the Manus assumption that only men enjoy playing with babies, the Toda proscription of almost all domestic work as too sacred for women, or the Arapesh insistence that women's heads are stronger than men's. . . . We found the Arapesh – both men and women – displaying a personality that, out of our historically limited preoccupations, we would call maternal in its parental aspects, and feminine in its sexual aspects. In marked contrast to these attitudes, we found among the Mundugumor that both men and women developed as ruthless, aggressive, positively sexed individuals . . . approximated to a personality type which we in our culture would only find in an undisciplined and very violent male. In the third tribe, the Tchambuli, we found a genuine reversal of the sex-attitudes of our own culture, with the woman the dominant, impersonal, managing partner, the man the less responsible and the emotionally dependent person.'[2]

I have given that quotation at some length because it is of the very greatest importance to realise that human nature is plastic, and can be shaped by society either for good or ill. If man were really an unregenerate and unimprovable brute, there would be no point in any cultural activities, or even in discussing the changes which could be made in society; what would be the purpose of making them if man himself became no better because of them? But such pessimism is unjustified; even those qualities of man which we usually consider the most spiritual, his unselfishness, his striving, his love for his fellow-men, can be altered and improved by changes in the social system in which he lives.

The present is one of those periods when society is being rapidly and profoundly altered. There are probably few people in Europe who feel that they fit in well to our system, with its wars, unemployment and general chaos; and it is obvious in all directions that changes are occurring. The development of society may take place in a more or less haphazard way, as a result of a multitude of small adjustments to particular evils, uncoordinated by any general point of view. But it would be more sensible if the people who find that they have outgrown the ways of life which society has tried to force on them decided first of all to discover what sort of people they wished to be, and then worked out what sort of society would allow them to be like that.

It may seem at first sight that it is quite obvious what the desirable qualities in a man are. It is easy to run off a list of virtues – energy, tolerance, creativeness and so on. But when we do so, we are usually thinking of these qualities within our present social framework, with all the presuppositions which it involves. When the framework gets out of date, and the system starts misfiring, it becomes less obvious which are the desirable attributes. The recent systems of Germany and Russia have certainly chosen some which we have not particularly valued – a ruthless devotion to the state in Germany, for example, and an equally selfless dedication to a particular concept of human progress in Russia.

But the seeds of the new life are here among us; there is no need, even if it were possible, to start out from scratch to invent a new ideal type of man. As our society has changed, and the conventional virtues have been frustrated, like enterprise, or gone out of date, like thrift, or even become social dangers, like competitiveness – all this time, the conditions of a new life have been developing, and the virtues which would make it possible to extract the greatest benefit from those conditions have become clearer. The people who, above all others, are interested in charting the development of human possibilities are the writers and artists. The function of culture is not, as many think, to enrich human life with new conceptions which it draws from some mysterious and transcendental realm out of the reach of ordinary people. The true, or at least the most important, task of the cultural worker is to reveal to man the spiritual riches which would result from the full exploitation of the immediately practical possibilities. The culture of Greece laid bare the greatest heights to which a slave society could aspire; that of the Renaissance worked out the potentialities of the newly-discovered learning in the circumstances of its own period of history. It is an exaggeration of this point of view to claim that the material conditions of existence determine the nature of culture. But they do limit it. The state of technical knowledge and the balance of social forces give man his opportunity. It is up to him to make the best of it; and in this endeavour it is the writers and artists, the scientists and architects who act as scouting parties to explore the terrain in front of the main body of the advance. It is from present-day culture that we can learn what sort of being modern man should be, and it is only when we know this that we can decide what political and economic system we must try to create.

In the past, artists and writers were able to perform alone this prime function of culture, the adaptation of man's ideals to the material world in which he lived. At present that is no longer true; they need to be joined by the scientist. Until about two hundred years ago, the material conditions of daily life changed very slowly. The new potentialities of life, which gradually arose, and which it was the duty of cultural activity to reveal, depended on changes in the economic organisation of society; the change from city states to empires, from slave civilisations to feudalism, or from feudalism to trade and capitalism. It may be true, as one school of historians assert, that in many cases these transitions were ultimately due to technical advances in handling the raw materials of life; for instance, the breakdown of feudalism may have been connected with the recruiting of large labour forces for working mines or mills using water power. But these technical improvements can at most only have acted as the last straws, the triggers which released a sequence of alterations which affected the lives of many people who had no direct or immediate contact with them. The psychological and cultural problems made themselves felt not in terms of material techniques but in connection with the changes in economic status. The technical man had no special appreciation of them, and no special contribution to make to their solution.

Nowadays the whole setting of our problems is obviously quite different. Everyone recognises that the difficulties which are shaking loose the old framework of civilised life are the results of technological changes. These act in two ways. New inventions and new processes have rapidly changed the day-to-day life of ordinary people; cars, electric light, telephones, radio and so on have made a world in which an adult man of Shakespeare's day would have difficulty in surviving for a week. And other technical advances, which do not enter directly into the experience of most people, have necessitated very great changes in the economic structure of society, and thus affect everyone at second-hand. The increase in size of the productive units – factories, firms, corporations, etc. – which carry on the business of the world is perhaps partly due to the inherent processes of capitalist economy, but it is also something more than a merely economic phenomenon; it is dictated by modern technological methods, and would occur under any economic system in which these methods could be used. Man, unless he is willing to give up the advantages of mass production and large-scale operation, can never again be his own master in the way a medieval craftsman was. And if civilisation is to continue to advance, and if full use is to be made of the opportunities for a richer life which technical advances present, the scientist and technician must join with the artist and writer in thinking out what those possibilities are and bringing them to the notice of people in general.

Up to the present, the collaboration of scientists in the general cultural activities has been very flimsy. They have mostly been content not to challenge the verdict passed on science many years ago by the encrusted incumbents (Beaune from the neck up) of ancient Professorships: that science is 'stinks' and has nothing to tell the humanities. The general adoption of this valuation by the cultivated world was the penalty which science paid for being allowed inside the privileged circle of the Universities.

In the seventeenth century it did not compete with Classics, Theology and Metaphysics for academic veneration and the easy existence of an endowed Chair, but it made no pretence that it was indifferent to its effect on social life. The Royal Society, the most august of English scientific bodies, was founded in 1662 by a group of men who shared the belief expressed by Boyle in the words 'the good of mankind may be much increased by the naturalist's insight into the trades', and many of whom had earlier come together in a society, the Invisible College, which 'values no knowledge but as it hath a tendency to use'. They were not all scientists as we now understand the term, but included also men like Wren, the architect, and Samuel Pepys, now famous for his diary but then an important administrative official in the Admiralty. At rather earlier periods, the examples of Sir Thomas Browne and Leonardo da Vinci are two very different instances of the way in which scientific and cultural interests and achievements were combined in the same man. Nowadays it would be only slightly unfair to take as typical of the relations between science and the arts the reviews of the Royal Academy which appear each year in *Nature*, the most widely read English scientific journal; very meticulously, the reviewer notes whether the geology of the landscapes, and the anatomy of the young ladies, is sufficiently text-book.

This decay of the relations between science and culture, and the concomitant withdrawal of science into the purely technical sphere, is often supported by the specious and ambiguous argument that science is, and must be, ethically neutral. In its most trivial and individualistic sense, that scientists do not mind which way their experiments come out, this is clearly untrue; nearly all active scientists more or less passionately hope to be able to prove or disprove some particular theory on which they are working. But that is irrelevant to the function of science as a cultural force. It is much more important that scientists must be ready for their pet theories to turn out to be wrong. Science as a whole certainly cannot allow its judgment about facts to be distorted by ideas of what ought to be true, or what one may hope to be true.

It cannot, for instance, allow its estimate of the relative food values of animal and vegetable foodstuffs to be influenced by the ethical arguments for vegetarianism. But it is stultifying and misleading to state that science can merely measure the advantages of the different animal proteins over the vegetable ones, and must then leave the ethical question entirely on one side, to be decided by others. The food values of various kinds of nourishment are an essential part of the whole situation on which an ethical judgment has to be made, and a part which, without the aid of science, we should remain ignorant of. If, or perhaps one should say when, science discovers some alternative and equally simple method of producing the food mankind needs, we shall quite likely think it more suitable to give up the somewhat inelegant apparatus of stockyards and slaughter houses on which we depend at present.

The contribution which science has to make to ethics, quite apart from questioning its fundamental presuppositions, but merely by revealing facts which were previously unknown or commonly overlooked, is very much greater than is usually admitted. The adoption of methods of thought which are commonplaces in science would bring before the bar of ethical judgment whole groups of phenomena which do not appear there now. For instance, our ethical notions are fundamentally based on a system of individual responsibility for individual acts. The principle of statistical correlation between two sets of events, although accepted in scientific practice, is not usually felt to be ethically completely valid. If a man hits a baby on the head with a hammer, we prosecute him for cruelty or murder; but if he sells dirty milk and the infant sickness or death rate goes up, we merely fine him for contravening the health laws.

And the ethical point is taken even less seriously when the responsibility, as well as the results of the crime, falls on a statistical assemblage. The whole community of England and Wales kills 8,000 babies a year by failing to bring its infant mortality rate down to the level reached by Oslo as early as 1931, which would be perfectly feasible;[3] but few people seem to think this a crime.

Quite recently, a new problem has arisen as the greatest conundrum facing the ethical judgment of man – the problem of the atomic bomb. Here it is inescapable that scientists must play a large, if not a dominant, role in deciding how man's new powers should be incorporated into his social life. Their responsibility is very large merely because of their knowledge. Anyone endowed with the normal human ideas of right and wrong can see that the bomb should be used as little as possible – though it is pertinent to point out that it was the scientists, not the non-scientific men, who protested against its use at Hiroshima and Nagasaki before the Japanese had been warned. When it comes to drawing up detailed measures to prevent resort to the bomb, it is only men with considerable scientific training who can appreciate the effects of various courses of action. Any system of control will come into some sort of conflict with the ideas of nationalism in which, unfortunately, so many of man's deepest ethical beliefs are nowadays involved. It is only scientists who are in possession of the information and theoretical understanding which can make it possible to decide on a system of control which conflicts as little as possible with other legitimate social values. There is no possibility for the physicist to fail to recognise his responsibility; and no excuse for non-physicists to deny the paramount importance of his counsel. And all this is true even if the scientist accepts without question the system of ethical values current in his time and his society.

But the ethical implications of a scientific attitude go even farther than this. The maintenance of a scientific attitude does in fact imply the assertion of a certain ethical standard. The reason that this has been overlooked, or denied, is that the scientific attitude consists in the overruling of the more obvious emotions which might interfere with the unbiassed appraisal of the situation; and the old-fashioned psychology which made a sharp distinction between the 'faculties' of thinking and feeling seemed to lead to the conclusion that science must banish all feeling and thus all ethical judgment. With the recognition in more recent times that such a distinction is unjustified, that all acts involve both feeling and thought, it becomes theoretically impossible to deny that 'feeling' is an element in the scientific attitude. Observation of the behaviour of scientists in their corporate and professional capacity confirms this. Before the war there was a very remarkable agreement among scientists throughout the world that a system of thought such as Nazism is incompatible with the scientific temper and is, for that reason among others, to be ethically condemned. Expressions of this point of view can be found in all the general periodicals of scientists, such as the English *Nature* and the American *Science*, with the exception of course of the officially controlled press of the Nazi or Fascist countries; and in the latter, the assertion of the ethical consequences of the scientific attitude has, as is well known, been made by many individuals who have suffered for doing so.

It is time, in fact, that scientists become willing to state explicitly that the scientific attitude is as full of passion, as much a function of the whole man and not merely of an intellectual part of him, as any other approach to human action. It differs from them only in what it is trying to do. Instead of trying to earn more money, or to improve the condition of the working class, or to create visual beauty, a scientist tries to find how things work. The search for causal connections cannot be made merely by refusing to grind axes. Scientific imagination and insight do not automatically result when the mind is swept clean of preconceived notions and prejudices; their attainment is a positive achievement and not a merely negative one. And because this is true, scientists can and do pass ethical judgment on human behaviour; those things which are based on the scientific attitude, or encourage it, are good, those which stultify or deny it are to that extent bad.

Finally, a more debatable, a more philosophical, and a less immediately important point: the scientific outlook has its own appropriate intellectual approach to the fundamental problem of deciding which, of the numerous and varied ethical beliefs man has held, is the best. There is no space or necessity to discuss this matter in detail here.[4] Roughly, the argument is that men's ethical beliefs influence their actions, and that one can observe these effects, and thus form a scientific theory as to the functions which ethical ideas fulfil in human life, just as one can form a theory of what function foods fulfil. Such a theory, I think, would have as its main thesis that the most important function of men's ethical beliefs is to provide a powerful mechanism by which human evolution is carried forward. If that is so, those ethical ideas which are most satisfactory in helping man along the path of evolution could be judged to be 'the best', in exactly the same sense that 'the best foods' are those which most satisfactorily fulfil the needs of his normal growth and development. Before one can make any practical use of this philosophical theory, of course, one must be able to discover at least the main outlines of the course of evolution – a difficult task, and one which must be approached without any narrow prejudices to look only for the obvious material factors, but a task which is no more difficult than that required by any other way of considering the eternal problems of right and wrong.

In the recent past, although science has been commonly held to be ethically neutral, and unconcerned with politics and social affairs, the scientific spirit has in fact been making an important contribution to the development of general cultural ideas. The goals towards which a society is moving are not always, perhaps not ever, completely conscious and formulated even by the most far-sighted individuals. A movement as powerful as science has been in our civilisation is bound to affect, even if unconsciously and at second-hand, the outlook of all those concerned with any aspect of the society's culture. We shall find, in fact, that an examination of recent artistic movements reveals a number of close connections with the scientific attitude of mind. The most constructive artistic outlook of recent times is one which shares very many of the characteristics of the scientific mentality; so much so that in Nazi Germany it fell under the same ban. The best of modern art is compatible only with true science, and a bogus science requires a fake art to keep it company.

In the next three chapters I shall therefore do my best to exhibit the cultural influence of science at the present day. It has recently become fashionable, when discussing the social influence of science, to put the emphasis on earlier periods of history – the geometry of the Egyptians, the astronomy of the Babylonians, and the ballistic problems tackled by the Renaissance mathematicians. But history is a heap of data in which one can bring to light examples to prove almost any theory one wishes; and when one has found a nice example, only the most self-confident specialist has any means of judging whether it holds water. No amount of palaver about the largely hypothetical past is a substitute for an analysis of the present, which can be checked by direct experience.

The study of the present, rather than the past, relations between science and society is the more essential nowadays because they have changed so considerably in recent times. In the past science was a comparatively minor activity of man, and the problems it tackled, and the course of development it followed, were to a very large extent dictated to it by man's other social activities. The social control over science is still a fact; a very penetrating and comprehensive analysis of it has been published by Bernal in his *Social Function of Science.* But science is now no longer so passive; it has acquired a momentum and strength of its own. Far from being content merely to accept the problems society suggests to it, it finds that it must pose problems to society. The thesis I want to argue is that science is already a very potent social force, and that it has certain social requirements on whose satisfaction it must insist.

CHAPTER III

ART BETWEEN THE WARS

The majority of artists, and those who become most popular, express the feelings and ideas which belong to the recent past or the immediate present. The ordinary man recognises himself in their works, and likes it. But the artists who shape the world to come are those who see a bit further than other people, who point out new ways, and force their message across, to some extent at any rate, on to a lazy and unwilling public. They are the creators, who do something more than reflect their surroundings.

In the past, when events moved at a more decorous speed, it usually took a really original artist or poet about twenty or thirty years to become widely understood. Today, with the world in a wild stampede, the mediocre artists are far behind the general public, and the creative ones far in front. The conditions of life with which they are dealing are not, at the most generous estimate, more than twenty years old; there has simply not been time enough to assimilate the problems of modern life, to create some æsthetic solution of them, and to get that solution generally accepted. So there are no acknowledged representative modern artists and poets; most of the men who come into question, and whose work deals with the world as it is now, are still young and still in the stage of being called cranks. The works which I shall discuss may seem queer, but at any rate they pass two of the tests of public esteem; the highbrow and the lowbrow. They would be, I think, considered important by the professional critics of modern culture; and they do, within the range of goods they represent, fetch high prices. Picasso probably commands more per square inch than any other painter ever did whilst still living. Neither of these criteria will satisfy the middlebrows, but as they are mainly concerned that things should not be too extreme, their opinion may be worth having about what is good art, but is not important when one is considering, as I wish to, in which direction art is going, dragging society behind it.

For a somewhat similar reason, I shall not discuss novels or the cinema. In some ways they are the most important arts of the present day, but they are not the most indicative. They are the arts which keep in closest touch with ordinary life. In novels the new ideas, whether original or borrowed, are worked in with the great mass of human characteristics which gives life its continuity; and although they are thereby reinforced and given body, they are, from another point of view, diluted and made less easy to recognise. If one compares a whole culture to a valley, the novels are the great rivers of the plain, on which the traffic flows; but the same slope of the land is more obvious up in the uncouth hills where the little streams, the poetry and painting, make a great clatter but gather no moss.

The meaning of a work of art is as difficult to describe as the expression of a human face or the song of a bird. The words of everyday prose are too crude to convey what the artist or the poet means; otherwise he would have used them. The enormous labour a creative worker puts into a painting or a poem is not done just for fun, fun though it may sometimes be; and the apparent strangeness which the result generally has for his contemporaries is not due to his perversity, but to the fact that he is trying to convey something which is new and therefore, since he has not had time to think it out fully, complex. But when one is considering, as we are, the importance of artistic movement for the social strivings of an epoch, one need not be bothered with the most troublesome of all questions about art. There is no need to consider how to judge the ultimate value of a work of art, or which are the best artists. The aspects of art or poetry which we shall be considering here are almost independent of æsthetic value. Some people – the believers in Art for Art's sake – would say that the value of a poem or picture is completely disconnected from its meaning; they claim that the only thing which is significant æsthetically is the mysterious magic for which we can give no recipe, but which we can recognise as the quality which differentiates poetry from verse or a picture from a mere representation. Personally, I think this goes too far; I do not believe that a man can be a really great artist unless his pictures contain not only the purely æsthetic magic, but also some attitude, of a generous and large-scale kind, to important aspects of human life. But certainly one cannot judge the value of artists merely by their social outlook; and that is why a discussion of the influence of artists on their contemporaries must largely disregard their ultimate æsthetic value as later generations recognise it. If artists were nothing but creators of visual magic, we could neglect them as an influence on the direction in which society develops; but they are not; the æsthetic power is a bait which, when accepted, carries with it all kinds of comments on the life by which the artist is surrounded and which intrudes itself, whether he will or not, into his work. For later generations, this may be comparatively unimportant, but for those who live at the same time and place, it is this intrusive 'meaning' which constitutes much of the artist's importance.

From the social point of view, the importance of an artist is not only measured by the number of people, interested in art, who think him good, but also by the effect he has on the artistic surroundings of the ordinary man who is under the mistaken impression that he never looks at a work of art from one year's end to another. The latter category is very important. Everybody, whether or not he ever goes to picture galleries or reads highbrow literature, is continually seeing posters and films, and reading the prose of advertisements and newspapers. A poster or advertisement of ten years ago looks obviously out of date today (except for patent medicine advertisements, which are nearly always designed to look ten years out of date anyhow, since they are aimed at older people with old-fashioned ideas). This passé appearance of the popular art of a few years ago shows how rapidly public taste changes. And it does not change because of some occult law of nature, but as a result, finally, of the creative work of highbrow artists. The man in the Underground may never have seen a picture by Picasso, and if he did, might dislike it; but he would not regard it with quite the same blank incomprehension or shocked hatred as did his father, who had never sat opposite a watered-down Picasso advising him to Go By Underground – It's Quicker.

The artists themselves know very well that they are the people who bring about the changes in popular taste. Struggling as they usually are to make both ends meet, they must often be tempted to give up the trouble of being a jump ahead of the rest of the world and to start earning the fat salaries of the people who adapt last year's artistic ideas to commercial purposes. If the only result of their creative activity was to alter the general lay-out of advertisements or the prose style of casual writers, there would be no important reason why they should go on with it. But most of them realise more or less consciously that these things are merely symbols of more important changes. Writing and pictures are methods of conveying ideas or feelings, and one cannot alter their character without altering the thoughts behind them. The sonorous and weighty sentences of a Victorian, with the adjectives descending from Olympian heights in neat squadrons of three, and the subordinate clauses rolling irrefutably after each other like the waves of a regular periodic function, belong to an age when one expected people to believe what one had to say. Nowadays we feel we ought to provide sixpennyworth of cracks, and you can laugh at the rest gratis if you want to. Measured utterances have disappeared from present-day use because our thought, more turbulent and contradictory, more ready to apprehend something of which it had been previously unaware, does not fit into that form. Their disappearance is deeply connected with our change to a society in which the comfortably-off have orange juice before breakfast instead of family prayers.

It is usual to quote Plato as an authority for the view that the ideas of poets and painters eventually get across to their contemporaries and modify the basic assumptions on which their way of life is founded. It is more important to find that the artists themselves express the same thing. Although Picasso has devoted his whole life to painting furiously in his studio without apparently caring in the slightest what the world thinks of the results, he has probably had as great an influence as any living man on the kind of approach to the world which the Englishman, Frenchman or American finds interesting. And,

i would
suggest that certain ideas gestures
rhymes, like Gillette Razor Blades
having been used and reused
to the mystical moment of dullness emphatically are
Not To Be Resharpened. (Case in point

if we believe these gently O sweetly
melancholy trillers amid the thrillers
these crespuscular violinists among my and your
skyscrapers – Helen & Cleopatra were Just Too Lovely,
The Snail's On The Thorn enter Morn and God's
In His andsoforth

(do you get me?) according
to such supposedly indigenous
throstles Art is O World O Life
a formula: example, Turn Your Shirttails Into

Drawers and If It Isn't an Eastman It Isn't A
Kodak therefore my friends let
us now sing each and all fortissimo A –
mer
i

ca, I
love
You. And there's a
hun-dred-mil-lion-others, like
all of you successfully if
delicately gelded (or spaded)
gentlemen (and ladies) – pretty

littleliverpill –
hearted-Nujolneeding-There's-A-Reason
americans (who tensetendoned and with
upward vacant eyes, painfully
perpetually crouched, quivering, upon the
sternly allotted sandpile
– how silently
emit a tiny violetflavoured nuisance: Odor?

ono.
comes out lies a ribbon likes flat on the brush.[8]

This bawling-out of the universe, though it might be fun while it lasted, could not of course get anyone anywhere. Eliot, after a hard struggle, developed out of his original disillusionment into a profound poet, though one who relied too much on the Church; but Cummings could only continue to develop his typographical extravagances into a sort of word-making-and-word-taking game which is almost completely incomprehensible except for a vaguely Fascist flavour and occasional sparks of obscenity.

There were other artistic movements which seemed at first to be just as purely destructive, but which have since turned out to provide a basis for new types of constructive activity which seem to lead somewhere. Just before the 1914 war, some painters in Paris had started the cubist movement, which began by analysing the painted representation of an object into a set of simple interesecting planes. It was a highly technical affair, as incomprehensible to anyone but professional painters as relativity mathematics is to the non-mathematician; though since painters have to live by selling their pictures to the public, there was a great pretence that everyone could follow what they were doing. The influence of cubism on ordinary ways of thought was, in spite of its technicality, as profound as that of the relativity theory. Painters interested in cubism, and people who looked at cubist pictures or imitations of them, did not regard everyday objects in the everyday way, as things existing in a network of connections with other things, and associations with other ideas. A chair from the Cubist point of view was not something you sat on, and of a particular kind such as you are likely to find in a kitchen or a drawing-room or wherever it might be; it was just a thing of definite shape, whose geometrical properties could be analysed, for purposes of putting it on to canvas, in a particular way, quite independently of people or rooms or anything else. Such a restricted and technical way of looking at things was soon carried to its limit and had to be amplified by adding something more interesting. Its importance was that it emphasised in painting the general scientific movement to analyse the world, concentrating on some aspects of it and leaving out the rest.

The part which it was generally agreed to leave out was the mass of hackneyed associations which Eliot, Cummings and the others fulminated against. But the poets were much less radical than the painters. Cummings's apparently hard-boiled line gets pretty blurry when you follow it up. He attacked the sentimentality of other people merely in order to substitute his own: for instance,

touching you i say (it being Spring
and night) 'let us go a very little beyond
the last road – there's something to be found'

and smiling you answer everything
turns into something else, and slips away . . .
(these leaves are Thingish with moondrool
and i'm ever so very little afraid).[9]

And Eliot in his earlier and most influential poems put, in the place of the conventional ideas which he despised, a mixture of metaphors which emphasised the more depressing aspects of things, and associations with parts of culture which were too highbrow to have been contaminated by the common herd:

The pain of living and the drug of dreams
Curl up the small soul in the window seat
Behind the *Encyclopædia Britannica*.[10]

The painters, on the other hand, analysed and dissected the objects they were supposed to be painting, and pared away the usual associations to such an extent that they found themselves with nothing left. Much of the writing by modern painters or their friends has been devoted to discussing what has happened to the objects which they used to portray. For instance, an article written just before the war by one of the best young English painters, John Piper, was called 'Lost, A Valuable Object' and in it he said:[11] 'The one thing neither of them (of the two kinds of painters he is discussing) would dream of painting is a tree standing in a field. For the tree standing in the field has practically no meaning at the moment for the painter. It is an ideal; not a reality.' The artists of course regretted their inability to think of anything to paint. John Piper finishes his article thus: 'It will be a good thing to get back to the tree in the field that everyone is working for. For it is certainly to be hoped that we shall get back to it as a fact, as a reality. As something more than an ideal.'

Most painters have failed to find any real tree in the field. There are two different directions in which they looked for it. One group discovered that if one isolates an object completely from all its normal associations, a whole set of other associations, which are usually more or less buried and unconscious, tend to come to the surface of the mind, often bringing with them the peculiar emotional intensity which belongs to other unconscious manifestations such as dreams. The surrealists, or superrealists, set themselves to explore this new world of obscure symbolism which lies at the boundaries between normal existence and dreams or madness. They invented several techniques for getting themselves into the necessary state of mind; they went in for self-hypnotism or automatic writing or, more simply, relied on the surprising effect of bringing incongruous objects together in unlikely situations, as in one of their famous examples which spoke of an umbrella finding itself with a sewing machine on an operating table. Paul Nash[12] has explained the method of surrealism by referring to the poet André Breton, who said that 'a statue in a street or some place where it would normally be found is just a statue, as it were, in its right mind; but a statue in a ditch or in the middle of a ploughed field is then an object in a state of surrealism. . . . It has, in fact, the quality of a dream image, when things are so often incongruous and slightly frightening in their relation to time or place.'

The surrealists claimed that by showing that art was merely a question of getting into an unmatter-of-fact frame of mind, they had made it possible for everyone to be an artist. Everyone can dream and have irrational notions. When critics told them that they were merely playing with nonsense, they replied that the critics were being spiteful because the basis of professional criticism had been destroyed. Surrealism, said Max Ernst,[13] 'has opened up a field of vision limited only by the mind's capacity for nervous excitement. It goes without saying that this has been a great blow to the critics, who are terrified to see the "author's" importance being reduced to a minimum and the conception of talent abolished.'

But what they had overlooked, of course, is that although everyone can dream and play the lunatic, very few people want to do so all the time. Surrealism has always been one of the standard brands of English humour; one finds it in ballads, in Shakespeare and in nonsense writers such as Lear; *Alice in Wonderland* is probably the best surrealist piece of literature so far. One of its most interesting present-day exponents is the American humorist, Thurber. But these writers keep surrealism in its place. The modern group who call themselves Surrealists with a capital S fail because they try to use surrealism as a complete and satisfactory philosophy.

A new way of looking at the world which depends on getting one's self to the borderline of madness is not a practical proposition as a general rule of life for the world at large. As a specialised activity carried on by a few people, it may produce some valuable new knowledge. In fact, in the more or less scientific hands of psychoanalysts it has certainly done so, and it is as an adjunct of this very important branch of scientific investigation that surrealism will develop in the future. As an influence on our general outlook on the world, the most that can be said for it is something else which Max Ernst also claimed: 'We have no doubt that by yielding naturally to the business of subduing appearances and upsetting the relationships of "realities" it is helping, with a smile on its lips, to hasten the general crisis of consciousness due in our time.' But one would answer the smile with more cheerfulness if, instead of merely hastening the crisis, surrealism had been able to do something about solving it. As an attempt to find new objects worthy of the attention of a busy and intelligent world, it was too fairy-like to be a success.

The other main direction taken by modern painting, towards what is known as abstract art, has also failed to lead to a satisfactory 'tree in the field'. In fact, the painters who worked along this line have hardly as yet tried to come to terms with ordinary objects. They were interested in the relations between shapes and colours, in the effects which can be got by mere meaningless patterns and contrasts. This again is obviously a highly specialised and technical painter's affair, which one would at first sight think had nothing to do with ordinary life. But actually it has had a considerable amount to do with it. Abstract and formal pattern, of the kind the painters were interested in, is one of the commonest things in daily life, although one is usually hardly aware of it. Every manufactured article is made in some definite shape; chairs, typewriters, ashtrays are made in shapes suitable for the style of some particular age or way of life; the design of a page of newspaper, or the heading of a letter, reflects an attitude. And more than reflects it; these everyday appearances, that the user scarcely notices, but which somebody has thought out and put together, have a great effect on the people who handle them. Unconsciously their mind becomes attuned to a style, and something of the feeling of the designers gets across to the unwitting user.

The technical researches of the abstract artists, passing through the hands of the commercial designers, who somewhat messed them about, have undoubtedly had a great effect on the appearance of our man-made world.

It is very significant that nearly all the abstractionists were attracted to much the same kind of design. It was not flowery and ornamented, nor fluid and lyrical; it was a rather hard and severe art, sometimes clean and elegant, as in the pictures of Ben Nicholson and Helion, sometimes with Leger's rather brutal swagger, always depending on the subtle relations of simple things, circles, rectangles of different colours and textures, and firm, definite lines which might be drawn with a ruler or compass. The difference it made to everyday design is there for anyone to see in the coinage or postage stamps, particularly in the short-lived Edward VIII stamps, in which abstract simplicity was allowed to run away with itself.

Important as abstract painting has been as an influence on manufactured designs, it is not all that one could hope for as an art. Many of the artists themselves are not fully satisfied with it; as Piper pointed out in the quotation given above, one wants something more, one wants to get hold of real things again. An abstract picture is too indirect a way of approaching people; most spectators, unless they go in for that kind of thing, are conscious of getting from it only a very vague impression of a feeling, if that. But the difficulty is to put objects into paintings and keep the same freshness and freedom from moribund ideas, the same strength and clarity. When everyone is wondering just exactly where they stand, and just what things are important to them, what can the painter do with things except play the fool with them, like the surrealists; pretend it is still 1900, like the Royal Academy; or throw them out of the studio altogether like the abstractionists?

Circumstances conspired to force the artist to find an answer. It was not merely that the outbreak of the war showed the surrealists that lunacy (which they now saw all round them) is not enough, and exposed to the abstractionists the flimsiness of the vehicle in which they tried to convey their message. There were also potent bread-and-butter reasons bringing their painting to earth. Almost the only painters who were able to continue painting were those employed as war artists, recording the scenes of the strange world of war, or on the scheme for recording the monuments of England. Both these groups, in their official work, returned to representational painting; they were forced to look at trees in the field, and they had to discover how to do so. It was probably a good thing. Henry Moore's studies of coal-miners and of shelterers during the Blitz, Graham Sutherland's industrial scenes and sketches of burning towns, Piper's paintings of bombed ruins, and Paul Nash's of air fighting, were not only first-class art and first-class reporting; they also, if an outsider may hazard a guess, seem to have done these painters the valuable service of bringing them for a short time into inevadable contact with the major and obvious interests of their fellow men. It was a refreshing experience similar to that which the W.P.A. tried to provide for American artists in the middle 1930's, by setting the painters caught by the depression to doing 'social' jobs, such as painting murals for public buildings. But the war was much more genuinely big stuff than were the social ideas behind the Rooseveltian New Deal; and the relation of the English artist to it was more spontaneous and direct than that of the American to the somewhat incoherent economic experiments of his President. And the best war art of England was certainly better than anything W.P.A. produced.

Moore, Sutherland, Piper and Nash, the four artists mentioned above, are typical of the present leaders of British painting. Although several of them have allowed themselves at various times to be called abstractionists or surrealists, none of them was ever fully assimilated into the conventional schools. They have all in different ways made an effort to rescue painting from being merely a highly technical branch of æsthetics, only comprehensible to the specialist, and have gone far to make it once again an important element in the main stream of civilisation.

There are of course a number of quite considerable artists who have, all through the period between the wars, stood apart from the general movements which have just been described. Men like Matisse, and Derain, and even more Bonnard and de Segonzac have been doing work which contains all the elements of a complete picture, and is not reduced to one aspect of it like an abstract or a surrealist picture. Their work has not been a cultural force of the same importance as that of the more extreme artists, simply because the world was not in a state in which complete pictures, or complete poems, were possible. The time was ripe for destruction and debunking, and one could not avoid the necessity for it by closing one's eyes. Among the really talented artists, almost the only ones who did not become surrealists or abstractionists were some of the older men. Standing aside from the main lines of thought, they could not develop their own work very much; to a great extent, each of them repeated his own line for years on end. But they could and did keep alive the belief that a picture could have a more direct relation to the everyday world, and express something as it were in ordinary language instead of in mathematical symbols. They were the people who kept the factory wheels turning between the decision to scrap the old model and the time when the new design was ready to go into production.

Almost the only man who took part in the changes which occurred in painting since World War I, but who could go on painting the natural things around him without repeating himself or anyone else, is Picasso, one of the inventors of cubism. He gets away with it by an intense interest in the world and in painting, which is so direct and personal that he seems to have no remembrance of the old ideas which the world is sick and tired of, and by an invention so fertile that there is no need to call in traditional ideas to help out. He has tried all the styles and invented many of them. He has been cubist, abstract and surrealist. He has played all sorts of tricks with the world, tricks which often depend basically on scientific ideas; for instance, his habit of painting faces which are both profiles and full-face views at the same time would hardly have occurred to anyone before it became common knowledge that mathematics plays similar games with the co-ordinates of space and time. But he has never been content for long with pictures which are purely abstract or purely surrealist. He always comes back to painting the recognisable world, even if he does remodel it a good deal before he puts it down on canvas.

'To my misfortune,' he says, 'and probably to my delight, I

use things as my passions tell me to. What a miserable fate for a painter who adores blondes to have to stop himself putting them into a picture because they don't go with the basket of fruit! How awful for a painter who loathes apples to have to use them all the time because they go so nicely with the cloth! I put all the things I like into my pictures. So much the worse for the things – they just have to get on with it.'

So he 'distorts' things, as we say, to make them 'go'. And also to make them real. When Picasso paints a blonde, she is not like all the other blondes, Schoolgirl Complexion All Over, she is the genuine blonde article What Gentlemen Prefer. 'Academic training in beauty is a sham. We have been deceived; but so well deceived that we can scarcely get back even a shadow of the truth. The beauties of the Parthenon, Venuses, Nymphs, Narcissi, are so many lies. Art is not the application of a canon of beauty but what the instinct and the brain can conceive beyond any canon. When we love a woman we don't start measuring her limbs.' . . . 'We have infected the pictures in museums with all our mistakes, all our stupidities, all our poverty of spirit. We have turned them into petty and ridiculous things. We have been tied up to a fiction, instead of trying to sense what inner life there was in the man who painted them. There ought to be an absolute dictatorship – a dictatorship of painters – a dictatorship of one painter – to suppress all those who have betrayed us, to suppress the tricksters, to suppress the means of betrayal, to suppress mannerisms, to suppress charm, to suppress history, to suppress a whole heap of other things. But common sense always gets away with it. Above all, let's have a revolution against that! The true dictator will always be got down by the dictatorship of common sense – and perhaps not.'[14] But really yes. It is the dictatorship of the common sense of science which will get away with it; the common sense which is not afraid to acknowledge the rights of love, that peculiar commotion of the glands of internal secretion, but believes in measuring a woman's limbs before making her a present of a pair of slippers; which turns a table into a mathematical function much odder than anything the cubists ever drew.

For an intense and vivid life – which is what Picasso is preaching and what his paintings express – is not enough in itself if it has no intellectual and spiritual focus around which to concentrate, and no social background out of which to grow. And not even Picasso has been able to go far towards supplying them. The same failure is still obvious in the newest school of poets. Older people often complain that the Second World War did not produce a crop of poets comparable to those of the first. The reason, surely, was that poets today are trying to tackle a more important and difficult task. The poems which became famous in 1914–1918 or just later were, after all, no more than fine expressions of rather immature thought – the adolescent and uncomprehending enthusiasm of Rupert Brooke or the cynical disillusionment of Siegfried Sassoon and E. E. Cummings. The best poets of that war, such as Wilfred Owen and Herbert Read, did not become known till much later. The recent war had been gathering for so long before it actually broke out on us, that all the early and easy reactions had been worked off on the Spanish episode – as thoroughly versified as anyone could wish for. When the major struggle started, poets knew that a major creative effort was demanded of them. It is significant that the main school which has, so far, emerged (and who call themselves rather grandiloquently the Apocalyptics) proclaim that their object is to teach men to 'live more and exist less; it will be militant against all narrow, shallow half-thoughts and backdoor sniggerings'. They welcome the 'variety and multiplicity of life'; faith, enthusiasm, fertility are their watchwords. But this insistence on vigour and intensity, while salutary enough and an essential preliminary to any important advance in civilisation, is not yet canalised by any definite line of thought. It remains consciously and explicitly anarchic; and anarchism, attractive enough in its insistence on the value of the individual, remains always an emasculated movement because of its failure to recognise that man's individuality cannot be separated from the society in which he lives.

CHAPTER IV

ART LOOKS TO SCIENCE

Wholehearted destruction and tentative reconstruction; that is how one can sum up cultural activities between the two wars. What I want to argue in this chapter is that the paramount influence behind both these phases has been science. From being a matter of mundane 'stinks', unworthy of the attention of the man of letters, of the scholar, of the philosopher, of all those whose business is to further the evolution of man's soul, scientific thought has become the pattern for the creative activity of our age, our only mode of transport through the rough seas in front of us.

I mean, of course, something much more important than a mere use by poets and painters of scientific subject matter. In poetry, in fact, there has not recently been any remarkable increase in the scientific metaphors or in the selection of scientific subjects. In the pioneering days of science in the seventeenth century, Donne made a fuller use of scientific notions than any important poet has attempted since.

> As the tree's sap doth seeke the root below
> In winter, in my winter now I goe . . .
>
> Eternall God, (for whom whoever dare
> Seeke new expressions, doe the Circle square) . . .
>
> At the round earth's imagin'd corners, blow
> Your trumpets, Angells, . . .
>
> But as in cutting up a man that's dead,
> The body will not last out, to have read
> On every part, and therefore men direct
> Their speech to parts, that are of most effect . . .

Few poets at the present day seem to have attended anatomy lectures, with demonstrations, or at least if they have they were not so impressed as Donne was.

The poets who have tried to use scientific imagery in recent times have often introduced it only in their more flippant works, such as Aldous Huxley's well-known *Fifth Philosopher's Song*, which begins:

> A million million spermatozoa,
>   All of them alive:
> Out of their cataclysm but one poor Noah
>   Dare hope to survive.[15]

or Auden's remark:

> Are her fond responses
> All-or-none reactions?[16]

If they have been serious, the science has usually proved indigestible:

> Courage. Weren't strips of heart culture seen
> Of late mating two periodicities?
> Could not Professor Charles Darwin
> Graft annual upon perennial trees?[17]

is not a very inspiring exhortation. It is obviously extremely difficult for the poet to decide just how scientifically sophisticated he can be. At one end of the scale one finds lines, like these from Empson's *High Dive*, which involve too specialised mathematics:

> A cry, a greenish hollow undulation
> Echoes slapping across the enclosed bathing-pool.
> It is irrotational; one potential function
> (Hollow, the cry of hounds) will give the rule.[18]

And at the other end, there is the too simple, such as this image from a poem by Spender describing two lovers who lie

> Arm locked in arm, head against head
> Whilst the nerves' implicit contacts
> Through the hidden cables spark:[19]

in which a somewhat crude reference to the physiology of nervous conduction merely suggests, to me at any rate, an incongruous picture of telephone connections in a hole in the street.

In fact, one would probably have to admit that if the poets are taken at their face value, there are at the present time a considerable number who have written apparently anti-scientific poems. Day Lewis, whose very colloquial idiom has enabled him to handle scientific ideas more adroitly than the authors mentioned above, is afraid that a scientific world would be rather nasty:

> Pasteurise mother's milk
> Spoon out the waters of comfort in kilogrammes,
> Let love be clinic, let creation's pulse
> Keep Greenwich time, guard creature
> Against creator, and breed your supermen!
> But not from me.[20]

And in a poem in which the poet defends himself against various enemies, the third enemy is the pretentious scientific philosopher, who claims:

> God is a proposition,
> And we that prove him are his priests, his chosen.[21]

But these are attacks on pseudo-science, on a doctrine which has forgotten the breadth of outlook and the humility in the face of facts which are an essential part of true science. They are poems which a scientist can enjoy and agree with; in fact, I shall in later chapters attempt to refute the same heresies myself, in a more pedestrian way. These poems which seem to be attacks on science are really defences of it; they use part of science's own resources to suppress its fifth column. That paradox explains why I feel justified in claiming John Crowe Ransom's *Persistent Explorer*[22] as one of my favourite scientific poems. Parts of it go as follows:

> The noise of water teased his literal ears
> Which heard the distant drumming and thus scored:
> Water is falling – it fell – therefore it roared.
> But he cried, That is more than water I hear.
> . . . .

But listen as he might, look fast or slow,
It was water, only water, tons, of it
Dropping into the gorge, and every bit
Was water – the insipid chemical H_2O.

. . . .

The sound was tremendous, but it was no voice
That spoke to him. The spectacle was grand
But it spelled him nothing, nothing, and
Forbade him whether to cower or rejoice.

. . . .

But there were many ways of living too,
And let his enemies gibe, but let them say
That he would throw this continent away
And seek another country – as he would do.

A beautiful description of the formation of a scientific hypothesis, its experimental confirmation, and the refusal of the scientist to accept that as the final end of everything he could discover.

As far as subject matter goes, painters have on the whole shown more sympathy for science than have poets. There are many pictures whose whole subject matter is more or less scientific. The dream images and unconscious associations used by the surrealists of course existed before they were scientifically studied, but it is only since psychologists became interested in them that the world in general has been willing to give them much attention, and it is doubtful whether surrealism would have been taken up as a 'movement' in recent years if Freud had not lived. 'For we surrealists,' says Salvador Dali,[23] 'as you will easily see if you pay us the slightest attention, are not exactly artists and neither are we exactly true men of science.'

The subject matter of most abstract pictures is even more clearly chosen from the objects well known to science. Some artists, such as Ben Nicholson and Mondrian, use mainly the simplest and most perfectly geometrical shapes, such as circles, straight lines and rectangles. Others use more complex scientific concepts, such as series of lines which define a curve or a surface which is technically known as their envelope; one sees them in the paintings of Erni and some of the recent drawings and carvings of Barbara Hepworth. It is significant that the English abstract artists invited a scientist, Bernal, to contribute to their International Survey of Constructive Art, 'Circle'. He pointed out two further uses made by artists of scientific thought; the use of the more subtle relations of symmetry, and the use of irregular shapes which are not arbitrary but which are defined by some algebraic function, so that they might be a graph of some imperceptible physical quality, such as a distribution of electric charge.

These are some of the uses of science as subject matter by artists. It is in more fundamental respects that the influence of science has been really important. In the first place, the whole of what has been called above the destructive activity of the last twenty years was an essential prerequisite to the creation of a scientific style. Science is essentially analytical. It can look at any phenomenon, from something as cold and empty of significance as the orientation of a molecule on a surface, to the bearing of a man whose head is bloody but unbowed. But it must get its items separated; anything subjected to its scrutiny must be isolated from the mush of general goings-on in which it is normally embedded, it must be defined, or if it cannot be formally defined, at least one must be able to indicate what exactly is the thing one is talking about, and what else is the fortuitous rag-tag-and-bobtail that happens to be cluttering it up at the moment. And, after the twenty or thirty centuries of culture which we consciously inherit, everything was pretty completely mixed up with everything else. We had, as Picasso said, infected the pictures in our museums with all our stupidities, all our mistakes. Can anyone still hear God Save the King or the Star-Spangled Banner as a tune? Each is covered so deep with associations that our ears can hardly disentangle it. If a young man falls in love, the chances are that he finds his imagination cluttered up with a tedious rigmarole of roses and moonlight and so on. It was from science that the advice came to sort out the welter into its component parts; to separate the sexual physiology from the botany and meteorology.

Moreover, the associations and prejudices which had clustered round each item of straightforward experience, and were smothering them as ivy smothers a tree, were derived from the old world which at the end of the First World War was already tottering under the impact of science, and must soon be altered out of all recognition by it. The focal points of our social life should have noble embodiments, said the pundits; a bank should be built in the classical style, a railway station perhaps in Gothic. But, answered commonsense, hacking away a few chunks of ivy, we are not living in the eighteenth century's imitation of classical Greece, nor do engine drivers necessarily aspire to reach Heaven by solitary contemplation and mortification of the flesh. Honour thy father and thy mother, said the fifth commandment; but scientific psychology showed that honour was hardly an adequate way of dealing with the complexities of family life. Love is a sacrament, the romantic poets cried; but science, leading off in the first round with that troublesome customer by cutting down most of the tree along with the ivy, said that love was a physiological corollary of a particular hormone balance, and Aldous Huxley's characters let it go at that.

> Had they deceived us
> Or deceived themselves, the quiet-voiced elders,
> Bequeathing us merely a receipt for deceit?[24]

asks Eliot in a poem published quite recently. The world was already saying 'Yes' twenty years ago. And loud in the chorus was the voice of science.

If the old meanings clustering round our daily life were dead; if ancestral wisdom was a fraud or no longer applied; if society was sick, and we had purged it to its bare bones, where were we to turn for new values, fresh food to bring the convalescent back to health? We have seen the surrealist solution. Objects peeled of the coatings of emotion with which generations of human experience have covered them become ghosts of themselves, and can live a phantasmal life hovering on the lunatic fringe of consciousness, producing in men the vacuous mental excitement of paranoia. The only other alternative to the grime of the outworn past is to consider things not as concretions of ether men's thoughts and feelings about them, but as agents which produce effects in the world. To consider them, in fact, scientifically. If the moon is a body giving off light of such and such a wavelength and such and such an intensity, there is no smear of sentimentality over its face. Light of that colour and that brightness can be significant and moving, not because we have been told that the Moon is the Goddess of Love, but perhaps because it abolishes difference of colour and fineness of detail from what we see; or because we experience it at the same time as silence and a relief from the necessity of doing something soon; or even possibly for some more obscure reason to do with the habits of life of our biological ancestors.

The scientific attitude to the world does not in the slightest deny the emotional effects produced on men by their experience; what it tries to do is to classify the mechanisms by which these effects were produced. Some will be more, some less and some very little, dependent on associations inherited from the culture of the past. If the more culturally dependent are rejected, what is left is an immediate effect, directly related to the circumstances of the actual experience, not fictitiously heightened by a plausible reference to some general but spurious theory. Consider, as an example of a modern love lyric, a verse by Auden:[25]

> Lay your sleeping head, my love,
> Human on my faithless arm;
> Time and fevers burn away
> Individual beauty from
> Thoughtful children, and the grave
> Proves the child ephemeral;
> But in my arms till break of day
> Let the living creature lie,
> Mortal, guilty, but to me
> The entirely beautiful.

In every phrase the poet is emphasising his refusal to consider the girl as anything but an ordinary girl, insisting on her normal humanity, and in the last line but one specifically rejecting, for himself, the concepts 'mortal, guilty' with their theological connotations; but no one would claim that because of this matter-of-factness the poem is lacking in feeling.

A similar desire to discard the conventional ornaments, to get along without stage properties, can be seen in much of the best recent poetry of all the Western countries. Consider, for instance, this by Paul Eluard:

> On ne peut me connaître
> Mieux que tu me connais

Tes yeux dans lesquels nous dormons
Tous les deux
On fait à mes lumières d'homme
Un sort meilleur qu'aux nuits du monde

Tes yeux dans lesquels je voyage
Ont donnè aux gestes des routes
Un sens dètachè de la terre

Dans tes yeux ceux qui nous révèlent
Notre solitude infinie
Ne sont plus ce qu'ils croyaient être

On ne peut te connaître
Mieux que je te connais.

This attitude which I have called the scientific might be described in other ways. It is the matter-of-fact as against the romantic, the objective as against the subjective, the empirical, the unprejudiced, the *ad hoc* as against the *a priori*. The emotional tone which goes with it is quite definite and quite complex, although at first sight, and to those brought up in a different mode of thought, it may seem emotionless and banal, and although it does in fact reject the standard emotional responses which inherited culture have made the most automatic and obvious. But, as I said in an earlier chapter, this very rejection of inessential qualities, which now seem shoddy and superseded, is not an easy task. It demands a certain fervour, and it carries with it its own excitement, an excitement identical in kind with that of the scientist on the track of a new and relevant concept. And experiences stand out again bright and fresh and clean, like dolls which a child unwraps from a box of dirty cottonwool packing. Their corners are not rounded off by being seen through a fog of culture; each one has its full individual character as a part of the causal system of events which make up the world. At least that is what a scientific way of regarding experience would aim at, and when it is, rarely and with difficulty, fully attained, the achievement is no more 'emotionless' than it must have been to Dalton when he reduced the untidy pile of facts about chemical composition to the law of constant proportions and the atomic theory; or to Mendel when he swept away a whole rubbish heap of nonsense about heredity and replaced it by his simple notion of the hereditary factor.

Definiteness and individuality of things, an interest in them as agents which produce effects, and an admiration of the elegance which results when the desired effect is exactly achieved with the minimum of fuss, these are the qualities of which science has in recent times been the main sponsor. It is because artists and writers since the last war have found in them the only way to advance that one is justified in saying that science is now in a position to become the leader of the humanities. Look at an abstract picture, by Leger or Nicholson or Mondrian; at the very first glance one is struck by the bright colours, the definite shapes and a controlled precision like that of a mathematical theorem. Even Picasso, with his much more total and less intellectual approach, writes that he sees things not as symbols, but with an almost innocent directness which would have been quite foreign to most periods of history but is not incongruous in our scientific age. 'I paint a window just as I look out of a window. If a window looks wrong in a picture open, I draw the curtain and shut it, just as I would in my own room. One must act in painting, as in life, directly. . . . We must not discriminate between things. Where things are concerned, there are no class distinctions.'[14] Again, listen to Eliot pointing out that the only way forward, even the only way to remain fully alive, is to deny the facile romantic ecstasy and to forget the traditional knowledge:

In order to arrive there,
To arrive where you are, to get from where you are not,
You must go by a way wherein there is no ecstasy.
In order to arrive at what you do not know
You must go by a way which is the way of ignorance.[26]

And the most considerable of the younger poets, Auden, has often expressed his debt to the scientific work of psychologists, as, for instance, in his fine poem in memory of Freud. In *Spain*, which dealt with the opening stages of the war, he describes the world as it should be after peace has been reached; and the main intellectual activity he mentions is scientific research:

Tomorrow, perhaps, the future: the research on fatigue
And the movements of packers; the gradual exploring of all the
Octaves of radiation.[27]

The dedication of a recent book of his is a statement that the world is made hideous by baseless power, but can, and will, be saved by knowledge:

> Every eye must weep alone
> Till I Will be overthrown.
>
> But I Will can be removed,
> Not having sense enough
> To guard against I Know.

Of all the artists, it is perhaps the architects who have realised most fully both the scientific character of the point of view to which they have come, and the existence of an essentially poetic element in scientific thought. It is not difficult to see why; their field of activity is half scientific. Just as painters naturally associated with anatomists at the time when human bodies were first being systematically dissected and anatomy was made into a science, so architects, nowadays, confronted with new building materials such as glass and reinforced concrete, find it necessary to keep in touch with physicists and engineers. The usages of traditional building were suitable for materials like brick, stone and crude concrete, which will resist thrust or compression, but which are easily pulled apart by tension; wood was the strongest material at its disposal for holding tensions. The chief technical problems which fascinated the architects using those materials were the attempt to cover large areas with a single roof, without the need of supporting pillars, and the attempt to make a wall which was strong enough to support the roof and at the same time contained as large an area of transparent light as possible. Suddenly the modern architect is presented with materials, steel for taking tensions, reinforced concrete which takes both tensions and compressions, which enable him perfectly simply to roof far larger spaces than any of his ancestors could do with their most elaborate fan vaultings and domes, and completely to abolish the problem of the larger window, by making, if he wishes, the whole wall of glass. No wonder he must become an applied scientist to discover just what are the potentialities of his new materials.

Just as the architect's materials are new, so are many of his problems. Most of the large buildings which have been designed recently have been factories, hospitals, storehouses, etc., buildings whose uses are so definite that the architect must take them into consideration. He must know not only what will be done in the building, but how it will be done, and why, so that he can think out the most convenient form of building for that particular purpose. At every turn, the materials he handles and the purposes he handles them for force the architect into contact with scientific knowledge and scientific ways of thinking.

Science influences architecture, as it does abstract painting, not only in its materials but more profoundly, though perhaps less directly, in its general outlook. Marcel Breuer, who in his work makes very imaginative use of new materials (he invented the steel furniture made from bent tubes) has said:[28] 'The basis of modern architecture, however, is not the new materials, nor even the new form, but the new mentality; that is to say, the view we take and the manner in which we judge our needs. Thus modern architecture would exist even without reinforced concrete, plywood and linoleum. It would exist even in stone, wood and brick.' 'The values (of the new æsthetic)' says another architect, J. L. Martin,[29] 'precision, economy, exact finish, are not merely the results of technical limitation. They are the product of artistic selection.' And the reason they are selected is that they fit in with the scientific spirit, the adoption of which is the basic revolution which has taken place in the world of architecture. The attitude of the modern architect, who asks of a building, say a railway station, What is it which goes on here? what is my building required to do, and what will be the effects in practical use of possible alternatives in design? is a fundamentally scientific attitude. It must have been a very subsidiary and easily overlooked attitude among the classical architects of the eighteenth century, for instance, who would often make a lavatory as large as a bedroom for the sake of the symmetry of their façade, or place the kitchen along with the servants' quarters, and away from the dining room, so as to express the cleavage between the upper and lower classes. It is not yet the generally accepted attitude; banks prefer looking respectable, disguised as Georgian houses, to experimenting to find the most convenient form, and the amount of convenience which a dweller in the new suburbs will sacrifice in order to look a little Tudor about the doorway, Moorish in the gables, or

Spanish Colonial in the veranda is at first sight almost unbelievable.

The cultural basis of English suburbia has recently been analysed in a very interesting book (*The Castles on the Ground*) by J. M. Richards – a thoroughly educated modern architect who feels sympathy with the tendencies he is discussing. Richards argues that one should look, not at individual suburban houses, still less at details of them, but at the suburban environment as a whole; and he says that if we do so, we shall find that it has real value as a charming world of fantasy, which the white collar office worker has spontaneously created. It must not be judged, says Richards, by the standards of the other two great classes of modern man – the productive man interested in output, or the consumer interested in quality: it is the world of the man concerned with distribution. Which may be true enough as far as it goes, but it does not go far enough. The suburban ideal, as Richards admits, is one of escape, and no one can be said to have mastered his position in life if his one desire is to escape from it. Moreover it is not even an original world of fantasy into which the suburban dweller escapes; it is a ragbag of bits of the metropolis in the corner cinema, and bits of the country in his lawns and shrubs; the turrets, loggias and verandas with which his house is embellished are reminiscences of all the architectural styles taught in the local schools. If this is, indeed, as Richards claims, the genuine people's art of today, we have to admit that we can neither solve our problems nor think of anything new.

The most important point about architecture is that it is in it, more than in any of the other arts, that the stupendous practical effects of scientific and artistic theories become most obvious. The modern architect must ask himself not only how to build the most convenient factory; he must ask himself how to house a man who wishes to live. Before he can find the answer he must discover what goes on in a living-house; what is living, actually? It is no longer farming, as it was when the Tudor house was designed – they have become museum pieces: it is no longer the life of the bewigged and noble lords for whom the English Classicists built palaces – they have been turned into schools; it is no longer even the respectable bourgeois existence for which the Georgian terraces were built, with their servants' quarters in the attics and basements – they have been cut up into flats. The designers who try to set the stage for present-day life must attempt to discover its essential features; and by the setting they provide for it they will in their turn go far to determine its character.

Architects have found only two ways of approaching this problem. On the one hand they suggest some sort of rehash of the past; Academy Georgian, Tudor watered down with H. and C. in every bedroom, or a pot-pourri of all the past whose flavour is so faint that it may pass as inoffensive even when inflated to the size of a skyscraper. But this is clearly no solution at all, simply a shifting about from foot to foot before one decides in which direction the future lies. The alternative is to turn on man himself the same scientific scrutiny that the architect already uses on factories. And the most compelling characteristic of living in recent times, as opposed to the life of the recent past, is its scepticism, its abandonment of tradition, its devaluation of authority as such. It is the same characteristic as that which we have called the scientific attitude of the architect and artist; of course the same, since the architect also is a modern man, a contemporary of the advertisement copywriter and the blacksmith out of work for the first time in six hundred years.

But the artist and the architect have a special responsibility. It is their duty to sum up and define a way of life. The lack of horses leaves the village blacksmith puzzled, and the advertisement writer has no time to determine whether there is anything which can be called truth. The architect who wished to build for a scientific and sceptical age had to, whether he liked or not, find out what was left when scepticism had done its worst.

The pundits would say that nothing was left; values, they said, are based on faith, or on ethical intuition or a whole host of other mysterious things, but at any rate not on science. Once call in question the accepted ideals, the ways of life and goals of striving sanctified by tradition, and there will be nothing to replace them; life will become empty and meaningless, and there will be no reason to prefer one setting for it to another. Nothing of the kind happened. The architects who considered human life from the sceptical scientific point of view were almost unanimous as to what things were valuable to it and what kinds of houses would be good to build. The only reason why anyone should be surprised at this is because the philosophers had nearly succeeded in making a corner in ethics, and persuading the rest of the world that they were the only people who knew what goodness really meant. But that is pure bluff. Goodness is a perfectly ordinary notion which comes into every field of experience, though things which are good in one field may be not so good in another, and one may be willing to leave the Absolute and Essential Good to the philosophers, since they, like everyone else, have never been able to get their hands on it. But obviously, in the world of typewriters, to take an example with which I am having some trouble at the moment, goodness means a high capacity for carrying out the functions proper to typewriters, namely making a certain set of symbols on paper. Every biologist who performs experiments with rats knows that a rat is an animal with certain behaviour and functions; a good rat is one in which those functions have been able to develop in their most definite and characteristic form, and conditions are good for rats in proportion as they allow this development to proceed completely and harmoniously, not inhibiting or exaggerating one part at the expense of another.

It may be objected that this is simply a commonsense view of goodness, and not in any way specifically scientific. But that is quite incorrect; it has only become a commonsense view recently, as the scientific habit of mind has become generally adopted. In the past, and in other societies, animals have been judged on quite other grounds, like the sacred cows of India. I do not know for what reasons people decided that black cats are good and bring luck. Nor do I know what was the ethical value of the horse, the noblest of man's helpers; nor just how wicked were the animals which were tried and executed for witchcraft in the Middle Ages; but I am sure that commonsense in those days was applying non-scientific criteria to things which we should now judge according to the outlook of science. If that outlook has become commonsense, that is all a scientist can ask for. But actually, although most people would be willing to admit that it is fairly good sense, it has not yet become common enough.

In the same way as the rat, man of today, even separated

from his traditional culture, still appears to the scientific architect as a being with a certain character, requiring houses of a kind which will allow, in fact encourage, that character to express itself. Dig him out from under the thatched roof of a fake rusticity; disinter him from the palm-girt halls of the Hotel Splendide; search him out in a back room (with curtained-off kitchenette and bath) in a mansion built for a solid and fecund family in Maida Vale, release him from a row of back-to-backs in Shoreditch, and he will be found to be a creature who thrives in light and air, provided he gets the efficient heating and sanitation which modern engineering can provide; who would like to be tidy, but cannot manage it without help; who can enliven a simple wall with pictures to suit his taste, but merely shuts his mind when confronted with the perpetual frown of someone else's idea of a cornice. The man who has escaped from the mixture of rabbit warren and gin palace that we call a city is a definite enough sort of person. It is for him that the modern architect works.

It is this normal citizen that the architectural profession in England must satisfy in performing the largest demand which has ever been made on conscious intellectual town-planning – not only the remodelling of most of our big cities after their mauling from the air, but in addition the building of twenty new towns, not as haphazard growths, but as coherently thought-out communities fit for present-day living, able to meet the competition of the traditional charm of the small country town. The prospects look good. Although in the big city architecture of office blocks and factories, America is far ahead of us, and in the design of individual modern houses we have probably no architects as good as the best Swedish, Swiss, French or American, when it comes to the laying out of a town which is neither a chaotic jumble nor an exercise in intellectual geometry, it seems quite likely that the young English town planners will not show up too badly on the world stage. Good town-planning, like gardening, is the art of making things grow naturally, healthily and in the right places; and gardening is something the English have always been good at.

CHAPTER V

SCIENCE'S FAILURE AND SUCCESS

All the cultural activities of our epoch have failed in their main function. Neither painting nor literature has been able to arrive at a point of view positive and definite enough to be worth even considering as a basis for a new society. They have been very useful; they have cleared away a lot of mess which everyone wanted to see the last of; they have indicated, rather hazily, the direction in which a new outlook on the world might be found; but they have not drawn the curtains and enabled us to look through on to a promised land.

Among the failures, one must also include science. In fact, its failure to realise and fulfil its social function is probably the most unfortunate of them all. It is, I have argued, actually the basis of the attitude to which the other cultural activities were trying to attain. The social consequences of the scientific habit of mind should have been a common topic of argument and criticism, a generally recognised subject for creative thinking, something of which first the leaders of contemporary culture, and then every educated person, would have heard and read and thought. Instead of that, the scientific world has often been, in its public and explicit expressions of opinion, unwilling to admit that its attitude has any relevance to social life, and it has hardly ever even dared to suggest that it has an important contribution to make.

It is only quite recently that the official leaders of science have begun timidly to approach the conclusion that science may have ethical consequences. Even now one gets the impression that the admission is made not so much from a firm intellectual conviction, but rather because the more generous spirits amongst them could not pass over the harnessing of science to war purposes without trying to find some mitigating prospect for the future. But at least it is significant and encouraging that both the last two Presidents of the Royal Society, Sir William Bragg and Sir Henry Dale, have publicly urged that scientists, because of their special knowledge, have special responsibilities. In America the problem was raised in a more urgent and less theoretical form, in connection with the atomic bomb. Who should decide whether it should be dropped on a living city? Who should decide whether it should continue to be kept secret? Normally the mechanism for the social regulation of enterprise is less developed in America than in Europe; it is the country of 'individualism' which still regards socialism with the shocked and fearful horror we got over some fifty years ago. But one could hardly feel satisfied to hand the bomb over to the wisdom even of General Motors or Fords. It demanded control by the whole of society. The question was, who should represent society? Many of the scientists felt that, since they were the only people who understood the bomb, they should at least have a considerable say in deciding how to use it, and that their views on the ethical difficulties were to be taken seriously.

This was, however, only a partial realisation of the point I am trying to make, which is not only the (perfectly true) statement that for the solution of some ethical problems a detailed scientific knowledge is of the greatest value, but also that the general scientific method of dealing with situations should affect our whole thinking about social problems. And until the last year or two, even such partial admissions have been rare; when such matters are discussed at all, the authoritative announcements have usually been distressingly vague, a mere moaning for π in the sky. Poetry and the arts, in their tentative approaches to science before the war, were wooing a frigid adolescent.

There are several reasons for this: the first, financial. Scientists, unlike artists and writers, do not earn their living in a free-lance, hand-to-mouth way, doing whatever interests them most provided they can pick up from somewhere enough to keep going. They hold definite salaried positions. Many of these positions are in industrial and manufacturing businesses, and are of course concerned with the particular activities of the firms who pay for them. The posts in Universities are less strictly tied to commercial interests, but even there the money has to be found from somewhere, and ultimately its source is in the business world which controls all large sums of money at the present time. The resources which a University has most fully under its control are usually earmarked for subjects which were of interest to the original donors of the funds many years ago, and the only place to find the means to support new work of another type is in the hands of other interests who hold the strings of a long purse. And the business world is not primarily interested in the investigation of social life, or in the formulation of what the scientific attitude has to contribute to it. Why should it be? Business-men are caught up in a war to the death with their rivals, and what they want from scientific research is something which will help them to keep their heads above water – results which pay. Even if a successful man of affairs rises above the struggle and can indulge his fancy, his world is the world of production and manufacture, the sciences he is likely to be interested in are the technical ones, chemistry and physics and their newer offshoots. He is more likely to found a professorship in something like surface chemistry or crystal physics; perhaps only a retired advertising magnate would be expected to be primarily interested in social psychology.

In practice, in England this financial starvation of the social aspects of science has been mitigated from only two sources. It turned out to be one of the minor blessings of an empire that it has natives in it, and that, in order to rule them satisfactorily, one has to know something about them. It is not too difficult to earn your living as a student of mankind provided you study the Bantu or the Melanesians or some other more or less primitive tribe; and the Briton irritated by American sneers about the Empire is sometimes tempted to retort that the only reason the settlers did not liquidate the entire indigenous population of North America was in order to leave some raw material for the anthropologists. This somewhat uncouth way of dealing with perhaps the most important problem of the time even has certain advantages. The social systems of the native races in our colonies are usually simpler than our own in many respects, and it would, perhaps, be easier for a scientist to analyse them first than to start straightaway on our own more familiar but less manageable way of life. But this is a theoretical point; on balance this approach to social problems is both trifling and beside the point. The study of native races is a small and not very impressive particular part of science, whereas the study of society as a whole should be one of its most active branches, attracting some of its best minds. Moreover the particular question which is of crucial importance at the present time, the effect of the scientific attitude on social aims and structure, cannot by the nature of things be studied in the wilds of Africa, where the sociologist is probably the only particle of science in sight.

The main other sources of supply for the English student of society have been some of the great American research foundations, particularly the Rockefeller Foundation. The whole of the American way of life has been built up in a more or less uninhabited wilderness within the last four hundred years, and there has been scarcely any time during that period at which some part of the country was not beginning from scratch to form itself from a collection of isolated pioneers into a community. It is no wonder that Americans are more interested in the processes of social change than are Europeans, the inheritors of a centuries-long tradition of culture. Thus we find the president of the Rockefeller Foundation in 1939 writing as follows, and not only writing but implementing his words to the comforting tune of two million dollars:[30]

> Democracy today needs the social scientists, both inside and outside the universities. It needs to free them to think with all possible penetration, wherever that thinking may lead. New ideas about human relations and institutional adjustment should be fully, honestly and hospitably analysed. Society should be most deeply concerned not with ridiculing failures or condemning those whose findings it does not approve, but with aiding that small minority of pioneers whose work in the social studies is reaching up to new levels of scientific achievement. Such persons are to be found in universities, in government and in private life. No greater contribution to the disinterested comprehension of today's issues could be made than by affording these able men and women full opportunity to make their work genuinely effective.

This paragraph puts its finger on some of the other influences which have kept science in general and social science in particular from having its full effect on human affairs. Ridiculing failures and condemning those whose findings it does not approve are pastimes which society has been only too fond of; and scientists themselves are not guiltless of sometimes joining in. Often the pressure is a direct reactionary one in the ordinary political sense. It is clear that there are a great many inefficiencies and failures in our economic system which are kept in being because they are to the advantage of powerful interests; and those interests are not going to be pleased with a scientist who points them out. Social science, in fact, is a dangerous profession; if you are tactful enough not to offend the mighty, you are not likely to be interesting enough to cut much ice. And if the scientist himself did not realise it, the officials to whom he owed his appointment certainly did. 'Our university administrators,' says Professor Lynd of Princeton,[31] 'are concerned in their enforced daily decisions with the short-run welfare of an institution, and this may not be viewed as synonymous with the long-run welfare of our American culture. To go ahead frankly into the enlarged opportunity confronting the social sciences invites trouble. Putting one's head into the lion's mouth to operate on a sore tooth has its manifest disadvantages.'

The situation would have been much easier if the sociologists had been able to count on the respect and support of other scientists, but they usually could not. Science, at the time when organised religion was its enemy, had signed away its rights to have views on the most general questions in return for freedom to put off its swaddling clothes. The older humanities, philosophy, classics, literature, felt that the bargain should still be kept, and many scientists, brought up on the basis of that arrangement, agreed with them. The point of view was particularly clearly put by the physiologist, Professor A. V. Hill,[32] in 1933 (it is not necessarily his view now; the fact that in 1939 he entered Parliament as Member for Cambridge University suggests that he has changed his mind).

> If scientific people are to be accorded the privilege of immunity and tolerance by civilised societies, however, they must observe the rules. These rules could not be better summarised than they were 270 years ago by Robert Hooke. . . . 'The business and design of the Royal Society is – To improve the knowledge of natural things, and all useful Arts, Manufactures, Mechanick practises, Engynes and Inventions by Experiment – (not

meddling with Divinity, Metaphysicks, Moralls, Politicks, Grammar, Rhetorick or Logick)'; . . . Not meddling with morals and politics; such, I would urge, is the normal condition of tolerance and immunity for scientific pursuits in a civilised State . . . science should remain aloof and detached, not from any sense of superiority, not from any indifference to the common welfare, but as a condition of complete intellectual honesty.

Perhaps scientists would be happier if it could be so; but it is an impossible dream. Hooke's list makes pathetic reading. Divinity, Metaphysicks, Moralls, Politicks, Grammar, Rhetorick, Logick—science has meddled with them as gently as a bomber raid with an ammunition dump. Hardly one concept of any of these subjects has survived intact from Hooke's day to this, and science has been one of the main causes of their transmutation. Its influence on such general topics cannot be avoided; so surely it might as well be open, and subject to debate and criticism by scientists themselves as well as by those who only know the subject at second-hand. If tolerance is all that we hope for, a civilised society will tolerate opinions on matters of importance and not only on those which are too technical for it to understand.

But there is no doubt that among scientists themselves the social sciences were, at least in England, though less so in America, considered not quite respectable. There are of course more solid grounds for that view than mere timidity. The social behaviour of man is an activity of the sort which science has always found most difficult to study. It cannot be simply analysed into component parts, so that one distinguishes clearly, for instance, between economics and politics, or between either of these and traditional custom. A society is in some ways a unity, and all its different aspects overlap on to one another, so that it is extremely difficult to draw any valid conclusions about any one part without studying them all. Somewhere, one feels, there ought to be a clue to what it is which holds together all the diverse facets of a culture, its religion and its games, its business morals and its sense of individual worth, its economic structure and its sexual ethics. The standard scientific methods of dealing with things were worked out for simpler situations; when applied to objects which are both complicated and highly organised and unified, they tend to yield a lot of isolated facts about details but to let slip the secret of the thing as a whole, which is what one is after. It is an orthodox and respectable scientific technique to measure things, if necessary to measure two or more things, and find how well the measurements fit. Sociologists did so in a big way, determining how the marriage rate varied with the wholesale cost of manufactured or agricultural products, or how the proportion of a wage spent on rent varies with the amount of wage received and so on. Very interesting many of these facts are; but they should be the mere foundation of a science; they are 'data', things given, but not things which have been received and used and turned into something by the creative imagination.

As one of the most prominent American sociologists put it:[33]

> We social scientists have great arrays of data;
>
> . . . data on production and distribution, but not the data which will enable us to say with assurance, as the experts dealing with such matters, how our economy can get into use all of the needed goods we are physically capable of producing:
>
> . . . data on past business cycles, but not the data that enabled us to foresee the great depression of 1929 even six months before it occurred:
>
> . . . data on labour problems, but not the data to provide an effective programme for solving the central problems of unemployment and of the widening class-cleavage between capital and labour:
>
> . . . legal data, but not the data to implement us to curb admittedly increasing lawlessness:
>
> . . . data on public administration, but not the data for a well co-ordinated programme with which to attack such central problems of American democracy as the fading meaning of 'citizenship' to the urban dweller and what Secretary Wallace has called the 'private ownership ot government' by business:
>
> . . . data on the irrationality of human behaviour and on the wide inequalities in intelligence, but not the data on how a culture can be made to operate democratically by and for such human components.

If this was the state in America, English science was certainly no better off. In fact, being further away from the sources of financial support provided by the American research foundations, even its collections of data on social questions were much less complete. Above all, they were even less related to the living reality of life in our civilisation. We had our censuses and economic statistics, less exhaustive than the American but still fairly thorough. But there was on this side of the Atlantic very little to correspond to the enormous mass of American data on questions of personal taste and interest, on such matters as the relation between a husband's profession and his views on whether a woman's place is the home, or on the kinds of activities with which the unemployed tried to occupy themselves during their enforced leisure.

Something was done to improve things during the war. The Wartime Social Survey was set up, and, having dropped the 'Wartime' from its title, still keeps going. Its activities had one great practical advantage over those of most of the American investigations; they were intended to provide data for immediate use in dealing with problems of major importance to society, matters such as reactions to being bombed, attitudes to food rationing and so on. Most previous social surveys had either been, as Lynd suggests, rather pointless exercises, or were related to the comparatively narrow interests of particular sales or advertising organisations. Quite recently the technique of social surveying began to be used in intimate contact with the widest possible projects of social creation. It is becoming recognised as an integral part, just as important as the mapping of physical resources, in the planning of new towns or the replanning of old ones. For instance, in Max Lock's plan for Middlesbrough, the grouping of the new town into units, the provision of schools, shopping centres and so on, was based on a study of what in fact are the effective neighbourhood units at present, for which they went outside it, for which activities people remained in their neighbourhood, and so on.[34]

Such work established an almost ideal set-up for one end of the spectrum of scientific research, namely its connection with practice. But social science will only prosper if the other, more abstract, end is also adequately catered for. The practical work is bound to throw up problems (the nature of the cohesion of social groups, for instance) which will demand special experimental studies of no immediate practical application. Such work is bound to be costly. There are still very few people engaged in it; and it is difficult to believe there will be many more until a more satisfactory method of financing it is found.

Descriptive surveys of human behaviour, important and

worthy enterprises though they may be, are not by themselves enough to furnish a full understanding of our civilisation. A society is an organised whole, and cannot be completely comprehended by the mere collection of facts about it. The task is one which demands that peculiar quality which we call imagination or intuition or insight; and the first imaginative attempts to break the complex down into elements which have some importance are likely to produce ideas which have a somewhat cranky, half-baked look. The language in which they are expressed is usually uncouth and full of newly invented jargon, feeling after the new ideas which the author cannot yet put down simply but must try to convey by implication rather than by precise definition. Older sciences which are pushing ahead from a well-secured base at the end of a long and well-charted road of previous discoveries may well be tempted to look down their noses at such guerilla parties. But the skirmishes are out on the track of social order, and they will get a satisfactory analysis for it in time. The most promising of the purely scientific attacks is coming from the anthropologists. Malinowski showed how the behaviour of some tribes hung together as a system functioning to keep them alive and their society in being. The Americans Benedict and Mead, and the Englishman Bateson, have widened the analysis to include not only physical behaviour but also the intellectual system of beliefs and theories, and the set of emotional attitudes and feelings which are part of life in the group.

We could have made good use of such ideas in relation to our own society in its present state of disintegration and change if they had been more fully developed. But they have not been; and the last of the reasons one may mention for this comparative neglect is the parallel, within science, of the very disintegration of society which social science might have enabled us to overcome. For as our civilisation has lost its unity and become a collection of individuals with innumerable different sets of beliefs and ideals, science in general has tended to become an enormous collection of details. Again, there is a certain degree of justification for this. Scientific knowledge and understanding is a communal achievement, the sum of a multitude of contributions from many different people. Any individual may feel a certain justifiable pride if he knows that he has added one brick to the structure. But in recent years, more I think than in the past, it has been commonly assumed that the discovery of one or two nuggets of knowledge is all that a scientist need attempt. So long as his facts are correct, and his hypothesis within its narrow limits does not lead to any contradictions, his duty was thought to be done. Any attempt to take a broader view of a complex problem as a whole, and to assess the importance of the various elements in it, seemed inappropriate in an age when there were few generally accepted views about major problems in any field of activity. It was orthodox and acceptable to discuss the relation between two or possibly three of the individual concepts of science, but anything more comprehensive than this tended to be thought slightly disreputable, mere speculation or word spinning.

In the words of the old cliché, in fact, scientists have tended to refuse to see the wood for the trees. There have been an army of bricklayers piling brick on brick, even plumbers setting up super W.C.s and heating and lighting engineers installing the most modern equipment; but they have all united to shoo the architect off the building site, and the edifice of knowledge is growing like a factory with a furnace too big for its boilers, its precision tools installed in a room with no lighting, and anyhow with no one who knows what it is supposed to manufacture.

Just as the character of the typical modern man became so disorganised that he could no longer be described by some adjective such as Quaker or Liberal or Catholic, which applied to all aspects of him, but had to be called something like insurance agent or business executive, which leaves most of his character out of the picture; so the ideal of the scientific world ceased to be the man of science as a complete and adult man, all of whose activities were permeated with the scientific attitude of mind, and became the specialist, the veterinary helminthologist, say, who knew all about one or two families of parasitic worms, but kept this knowledge quite separate from anything else that he might happen to be interested in. To give a scientific opinion meant to expound the consequences of some specialised field of knowledge. Responsible scientists, looking at their colleagues, saw the obvious fact that most specialists were quite unfitted to play an important part in the evolution of general culture; but far from acknowledging that this was a sign of science's failure, they accepted it almost with glee as an excuse which let them out of the necessity of thinking about wider issues. It tended to be forgotten that there are certain general characteristics of scientific knowledge as a whole, and that where the specialised field has been inadequately developed, as in the social sciences, there may still be some opinions which follow logically and necessarily from these fundamental features common to all scientific thought.

Here is a very pretty example of the current confusion between the two kinds of scientist, the technical expert and the man who thinks scientifically. One of the four or five most influential scientists in England recently wrote[35] in praise of a young American physicist who said to him 'If I want to express opinions on morals or politics I do so as a citizen, not as a scientist.' That is reasonable enough if scientist means technical expert, since very few scientists are experts on wage rates, rearmament policies, etc. The plausibility of the sentence when interpreted in this trivial sense undoubtedly tempts one to believe that it is also true when 'scientist' is given its full, wide meaning. The recommendation is then, put bluntly, that when considering morals or politics one should discard the scientific habits of thought; the search for unbiassed evidence, the checking of theory by practice, the interest in causal relations and the lack of interest in things for their own sakes. It is advice that very few scientists are likely to admit to following, if it is put squarely and unambiguously before them. The trouble is that it usually is not presented clearly, but, as in this case, is cloaked in all kinds of ambiguities and half-truths, which make it easy to slink round the corner muttering 'It has nothing to do with me.' Even the author who quoted it probably did not really mean it; at least he used it, not as one would expect, against people who allowed their science to influence their politics, but against those whose politics had in his opinion influenced their science, which is quite another story; and he is himself justly respected as one of the most active and conscientious scientists concerned with public affairs. At any rate, whichever of these senses the physicist's dictum is supposed to have, they all agree in stating that science has nothing to do with politics or morals; and you can interpret *that* in

1 Leger was painting pictures like this *Three Women* as long ago as the early twenties. Look at the posters on the next hoarding, or the book covers in the next shop window, and you will see his influence. You are quite likely to have selected your toothpaste because the design of its packing played up the hard purity that he put across. (*Museum of Modern Art, New York*)

2 A surrealist picture by Salvador Dali. *A Paranoic Head*, which, placed on its side, becomes a group of Negroes sitting round a hut: two entirely separate images whose conjunction induces only a feeling of irresponsible bemusement.

Blonde by Picasso. He has put in the profile and full-face both at once, presumably because he liked them both, and realised that they belonged together.

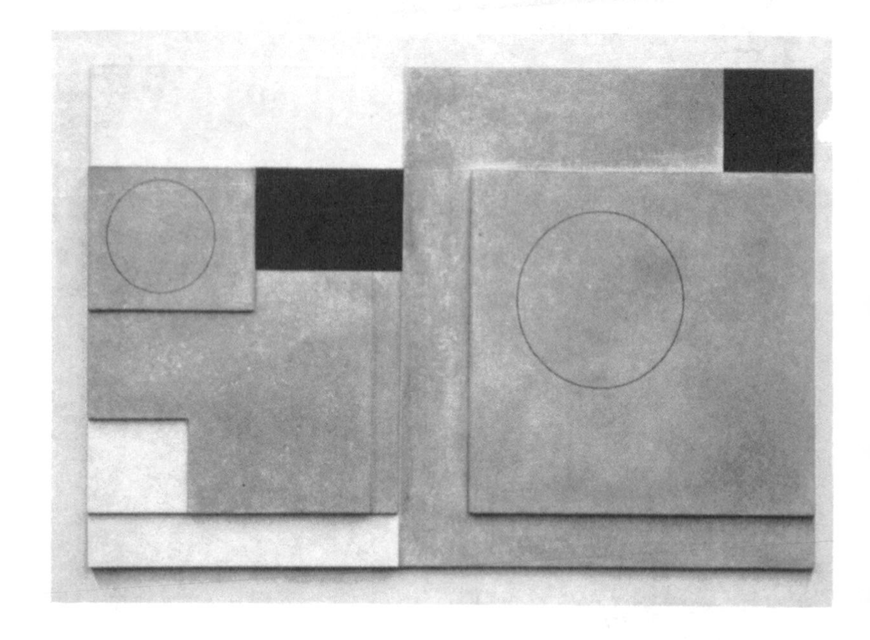

4 Ben Nicholson. A purely abstract bas-relief, which gets its effect of cheerful and undistracted elegance simply by the relations of a few plain surfaces. (*Museum of Modern Art, New York*)

5 ABOVE *Breakfast*, one of Picasso's simplified, distorted sketches which yet have all the essentials; an egg in a glass, a nice pat of butter and a roll – obviously a fine morning with no hangover.

6 OVERLEAF *The Control Room of the Central Headquarters, London, A.R.P.*, by John Piper. A picture by an abstract artist functioning as a war painter; and he has caught exactly the rather dingy administration of technical matters which is all there was to war in the early static period.

58
241
241

7 ABOVE A surrealist picture by Paul Klee, *The Twittering Machine*. The fantasy, as whimsical and ultimately as trivial as the Ascot hat of one of King Edward's lady friends, is exhibited with a profound originality of visual technique. (*Museum of Modern Art, New York*)

8 OVERLEAF *Holkham, Norfolk*, by John Piper, a picture in which an abstract artist has looked at the world again to find some real thing to paint; and has found something rather romantic.

9 The surrealist Magritte looks, matter-of-factly enough, at his meal; and his meal looks, fixedly, at him. (*Museum of Modern Art, New York*)

10 The Philips High Tension Generator for 'atom-smashing' in the Cavendish Laboratory, Cambridge. Here is the modern aesthetic at its source. A modern sculptor like Brancusi would have been proud of this, but his version would not have produced 2,000,000 volts.

11 A Victorian street, with bustle on the road, bustles on the pavement, and a forest of gas lamps in decorative cast iron.

2 A modern road. The 'furniture', beacons, 'Keep Left' signs, and so on, might have come out of a Leger painting. Besides giving us the trivial commands to stop and go, they are continually quietly reminding us that if we want to get anything done it is best not to say it with flowers, certainly not with cast-iron ones.

13 A municipal bathing place in Switzerland; the curved *roof* and the mushroom pillars are a brilliant technical use of a new material, concrete; but here the purpose of the building is simple human enjoyment.

14 Part of a Swiss housing scheme, which should convince anyone that the modern purity is not joyless or intolerant.

15 It was in Germany that some of the earliest and boldest experiments were made in architectural use of thin shells of concrete; it was difficult for the ordinary man, seeing his country's air force lodged in hangars of such daring design, to realise that the Nazi régime itself was not modern, but profoundly retrograde. This example, however, is of a more innocuous kind – a rum distillery in Mexico, designed by perhaps the greatest virtuoso in walking the tightrope between lightness and sufficient strength to stand up – Felix Candela.

several different ways, but there is one way at least, and that the most important, in which, as I have argued above, it is quite untrue.

In spite of this failure to realise that the scientific attitude has social implications, and in spite of the lack of development of the sciences which should deal with man's social behaviour, science has been a very potent cultural force. Its success within its own field has ensured that much, in a period when other branches of culture have produced very little whose value they can be certain of. What poets or artists or philosophers are there to match against the great scientists whose work has become of importance since the last war; Einstein, Rutherford, Bohr, Planck, Freud, Morgan and the rest? Some of their work may have been done before 1914, but it was in the between-wars period that it enlarged our basic ideas of the nature of space and time, of the material the world is made of, of how our minds work, and of the fundamental nature of living things. This was not an achievement of a few isolated men of genuis; in all these subjects the enormous advances in knowledge and understanding were aided by the efforts of a large number of people, some good, some not so good, but all together turning out a volume of radical but inescapable stuff with which the other cultural activities cannot begin to compete. No wonder poets and artists began to suck in a little scientific vapour with the air they breathed.

SCIENCE AND SOCIETY

CHAPTER VI

THE EMPTINESS OF FASCISM

For the first ten or so years after the First World War, the development of cultural activities in Germany was very similar to that in England and France. There was the same destruction and breaking down of accepted beliefs and customs, and the same attempts to build, from a scientific basis, something new to put in their place. The main difference was that in Germany both movements were more violent. The war did not cease for the Germans at the Armistice in 1918; the blockade against them was enforced for nearly a year longer, and there was fighting going on in Poland, Silesia and the Baltic countries for some time. Part of their country was occupied, and their economic system blew up into fragments in the great inflation which destroyed the whole structure of their society. No wonder that, even more than in other countries, people felt that they had no secure position in life, that all their preconceived beliefs and traditional values had failed them and must be got rid of; that liberty, equality and fraternity were moonshine; faith, hope and charity ended in the gutter; and even money didn't mean anything for more than a few days or hours at a time.

So their artists and writers were more virulently and crassly destructive than those in any other country. No pictures have ever been aimed at all previous ideas of beauty with more sardonic hate than the Merzbilde of Kurt Schwitters, made as mosaics by gumming together bits of dirty newspaper, bus tickets and so on. It would be difficult to find caricatures more devastating, and less relieved by humour, than the drawings of

George Gorsz. One of the great popular successes of about 1930, and a considerable work of art as well, was the *Three Ha'penny Opera*, modelled to some extent on our *Beggars' Opera*, but placed in a modern setting. Its pessimism was extreme and very complex. The highwaymen and their girls were transformed into a sort of fifth column of the ill-used; with all the cruelty, as well as the pity, that that implies; definitely 'on our side', but equally definitely knowing that they would be defeated.

What love poem could be more flat and hopeless than this by Kastner:[36]

> Du zürntest manchmal über meine Kühle.
> Ich muss dir sagen: Damals warst Du klug.
> Ich hatte stets die nämlichen Gefühle.
> Sie waren aber niemals stark genug.
>
> (You blazed up sometimes about my coolness.
> I can tell you: You had something there.
> I always had the relevant emotions.
> But never enough to turn a hair.)

On a higher plane, there was the surprising popularity of the poet Rilke. A great poet, too difficult to be widely read in ordinary times; but he expressed a pessimistic mysticism which suited the feeling of that time; for instance, in the lines:[37]

> Dir schien,
> weil du gewohnt warst an die andern Masse,
> es wäre nur für eine Weile; aber
> nun warst du in der Zeit, und Zeit ist lang.
> Und Zeit geht hin, und Zeit nimmt zu, und Zeit
> ist wie der Rückfall einer langen Krankheit.
>
> (You thought,
> because you were used to another standard,
> it was only for a while; but now
> you were caught in Time, and Time is long.
> And Time goes on, and Time mounts up, and Time
> is like the relapse of a long illness.)

There were also German movements which began to work out a new conception of life which would have its own values and could provide something by which a man could find life worth living. They accepted the work of the cultural demolition squads, and also tried to build something new on the cleared ground. They were daring and thorough, and in their time were in the very forefront of all the European groups which were trying to lay down the lines along which a new society could be formed. The most famous of the German groups, and the most influential, was known as the Bauhaus, or building-school (now enjoying a second lease of life in Chicago). For it was again the architects who, because their artistic creation is most nearly in contact with the lives of ordinary people, had been forced both to accept the fact that the traditional ways of life had led to disaster and to find some point of view which would provide a setting for the life which continued. The attitude with which they approached their work was very clearly the parallel, in the artistic field, of the mental attitude of the scientist in his field. Instead of dealing with things as objects full of meanings which have been given them by people in the past, and have been handed on to us by tradition, they looked at them with fresh eyes, neglecting their coating of acquired meanings, but analysing them for artistic purposes into their basic colours, textures and patterns. With these elements they started to make their new works of art.

They worked consciously for a scientific age, an age of mass production and machine finishes. The most important, or at least the most influential part of their output did not consist of pictures or works of art in the conventional sense, but of designs for things of practical use, crockery, saucepans, furniture, books, posters and so on. Their products were taken up and mass produced by German firms, and the style of their designs spread from there over the whole world. A very great deal of the more up-to-date parts of our surroundings look the way they do because of the work of the Bauhaus. The simple spherical hanging electric light fittings, wall papers which have a texture but no definite pattern, steel furniture, built-in cupboards, are some examples of things which they either invented or first popularised. If the objects among which a man spends his life have any influence on his outlook and feelings, the group of men who worked at the Bauhaus are some of the most important artists there have been since the last war.

But they lost in their own country and they have not yet won anywhere. The new way of life which they were pioneering was too slow in coming. The old German culture had been a rather tightly organised one; people knew their place and what was expected of them, and they had only to behave themselves to be free to drink their beer and listen to their concerts and operas and enjoy themselves. The disintegration of society and the disappearance of all generally accepted rules of what was fitting and right were probably harder for the Germans to bear than for any other nation of Europe; they were brought up to do their recognised duty and they could not bear the effort of thinking out for themselves what it was. When the great depression of the 1930's came, their civilisation was more completely disorganised and out of gear than anyone else's. There are good pictures of it in the novels of the time, Kästner's *Fabian* and Döblin's *Berlin Alexanderplatz*, for instance. The whole place was chaos; and people were ready to accept any cure so long as it came quickly.

The reasons why it was the Nazi cure which was accepted were of course complex, as reasons always are in social affairs. One of the main causes was that in Germany, as in the rest of the world to this day, the forces which were defining a new outlook of a scientific type had failed to realise their unity and had been unable to get themselves across to people with sufficient clarity and explicitness to constitute a strong volume of powerfully held belief. In particular, the scientists had as a whole stood apart from the general cultural movement which was ultimately based on their unconscious influence. They carried on with the lazy, timid belief that science is concerned with a variety of out-of-the-way phenomena among electrons and insects, but not with the daily life of man; and they still had the dangerous illusion that the other side made the same mistake. Then the leaders of the political parties who could most easily have adopted the scientific attitude, the Social Democrats and Communists, were undecided and weak, while their opponents, who stood for beliefs incompatible with science, were ruthless and daring, and reaped the advantages of money (not that money was a very definite thing in post-inflation Germany) and a tradition of ruling (which should have counted for nothing at a time when all traditions were garbled).

The reactionary political forces would have been comparatively powerless if they had not been able to find some other system of thought, which equally with the scientific was a break with the past, and which could be pushed as an alternative way out of Germany's confusion. Such an alternative existed in Germany much more strongly than in any other European countries, in the form of the youth movements. It was a mixture of the ideas which were running through these movements with old-fashioned nationalism and militarism which produced the witches' brew of Hitlerism, and it was the elements derived from the youth movements which gave the real kick to it.

The influence of these pre-Hitlerian youth movements is too often forgotten at the present time, but I believe it is impossible to understand the peculiar nature of the Nazi system of beliefs without remembering them. They were groups of young men, and sometimes women, who went in for hiking, camping and sports generally, rather like boy scouts. But they were young people profoundly disillusioned with the world, and the knots they played with were tied not in string but in their own personalities. They rejected the squalid inflation-world they lived in and went out into the woods to look for something else. They were young and inexperienced. They found nothing of what they were looking for: nothing at all; but they made do with that.

It is the fundamental fact about Nazism that at the basis of it, where there should be the foundations on which the whole structure of beliefs rests, there is nothing at all. It is just talking to keep one's courage up, holding one's self up by one's socks. The sort of thing the youth movements came back from the woods talking about was, to quote one of their few English representatives, 'a night-time religion – that elemental unreason from which wisdom is absorbed as nurture is sucked by a tree-root from the soil'. That was the kind of wind in the trees which gave these empty young men the illusion that they had filled their bellies with solid meat.

Nazi intellectuals were later to write almost exactly the same kind of nonsense and describe it as the basis of their whole outlook.[38] 'Blood and soil, as the fundamental forces of life, are, however, the symbols of the national-political point of view and of the heroic style of life. By them the ground is prepared for a new form of education. . . . What does blood mean to us?

We cannot rest satisfied with the teachings of physics, chemistry or medicine. From the earliest dawn of the race, this blood, this shadowy stream of life, has had a symbolic significance and leads us into the realm of metaphysics. Blood is the builder of the body, and also the source of the spirit of the race. In blood lurks our ancestral inheritance, in blood is embodied the race, from blood arise the character and destiny of man; blood is to man the hidden undercurrent, the symbol of the current of life from which man can arise and ascend to the regions of light, of spirit and of knowledge.'

Nothing but words, chatter. The author does not mean blood in the ordinary sense; he specifically rejects the ordinary means of knowledge about it, and he probably knows as well as anyone that our ancestral inheritance is no more, and no less, in our blood than in our kidneys, that our character arises from our blood much less than from our pituitary glands. Blood is just an emotional word to him; it doesn't mean anything, but it makes you feel good.

But words are powerful things. The irrational nonsense turned out by Rosenberg and the other Nazi theorists did produce a comfortable feeling in many puzzled and unhappy Germans. It was so sweet a jam, and was pushed down their throats with such persistence, that they swallowed the most amazingly unpleasant pills along with it. They would not have taken it, probably, if there had not been very powerful groups forcing it on them. But similarly the big industrialists and bankers who provided Hitler's war chest could not have steered the development of society in the way they wanted it to go if they had not been able to find some fundamental attitude which they could use as a vehicle to put over their programme.

Money power cannot bring about changes in social structure directly; it must find some method of translating itself into emotional attitudes and intellectual values. The importance of this step is obvious from a comparison of Germany with Spain; Hitler, who made a psychological attack, entirely destroyed the most powerful labour movement in Europe with very little bloodshed, and not only enslaved the workers in his machine, but persuaded the majority of them to like it, while Franco and his allies, making a direct military attack with no preliminary psychological groundbait, were held at bay for two years by a completely new and only half-organised democracy, and finally achieved a much less complete victory. The purely economic explanations of Hitlerism, which go no further than to classify it as a form of monopoly capitalism, are unduly simplified. They are dangerous because they obscure the fact that the stage of translating power into socially effective dogma is one of the few occasions when the privileged few have to descend to much the same level as everybody else, and can be met on more or less equal terms. They may still have some advantage in the means of propaganda at their disposal, but the goods they have to sell are so much less inherently attractive to the prospective purchasers that they should be in the position of needing all the help they can get and then some more.

All the other fascist philosophies have the same fundamental baselessness as the Nazi system. If one digs down to the buried chest where they keep their ultimate beliefs, and looks inside it, one feels like a liaison between Mr Jorrocks and Old Mother Hubbard; it is hellish dark and smells of cheese, and empty at that. Italian Fascism, according to Mussolini,[39] 'conceives of life as a struggle, considering that it behoves man to conquer for himself that life truly worthy of him.' But if one asks what is the life truly worthy of man, one does not get much help. 'The world seen through Fascism is not this material world which appears on the surface, in which man is an individual separated from all others and standing by himself, and in which he is governed by a natural law that makes him instinctively live a life of selfish and momentary pleasure. The man of Fascism is an individual who is nation and fatherland, which is a moral law, binding together individuals and the generations into a tradition and a mission, suppressing the instinct for a life enclosed within the brief round of pleasure in order to restore within duty a higher life free from the limits of time and space; a life in which the individual, through the denial of himself, through the sacrifice of his own private interests, through death itself, realises that completely spiritual existence in which his value as a man lies.'

Which means, so far as I can see, no more than that man is naturally bad, on the material plane, but can become spiritually valuable if he does his duty. But again, duty to what? Presumably to Mussolini, as no higher authority is specified. The appeal to spirituality means exactly nothing except a command to the Fascist to be obedient.

The Nazi theories did not remain a mere bundle of nonsense. In the first place, they achieved what the German people were asking for. They produced a coherent society; not one, perhaps, in which the average German was very happy, but at least one which could be understood, which had a fairly definite character and aims; a society in which, if one went with the stream, one did not need to feel lost, at any rate for a time. And, secondly, although the foundations of the society were profoundly irrational, there was no need for all the superstructure to be so too. There were many Nazi arguments, subsidiary to its main theory but still important enough, which were perfectly sensible and probably true, at least in their negative aspects: for instance, their criticism of the structure of Europe, that a hotchpotch of small independent nations is simply not feasible in the space at the present date in history, with a technical civilisation as highly developed as ours.

Again, in some technical spheres, Germany under the Nazis showed the same bold competence which always has been one of her characteristics. Pre-Nazi Germany had done some very fine work in building housing schemes and garden suburbs; although the Nazis cut down on this, they kept some of the projects going, and managed to persuade many people that they were responsible for the rest, which was quite untrue. But in subjects which fitted in with their social programme, they have often really been strikingly efficient. For instance, some of their aeroplane hangars are extremely beautiful buildings, making the most daring and successful use of new materials like shell-concrete in a way which gives a thrill to anyone interested in technical achievements. The great auto-roads, with their viaducts, bridges, and underpasses, were wonderful feats of engineering, flowing across the countryside with a magnificence and sweep which can leave no one unmoved. Germans, seeing these things and being proud of them, felt that the régime which produces them must be worth something. It is difficult to remember always that technical efficiency is not enough by itself, and that it can be used just as well or better to give man a richer and fuller life as for purposes which ultimately neglect his welfare and serve only the glory of the State.

The excellence of some of their technique cannot obscure the fundamental opposition between the Nazi ideas and the scientific attitude to society. One of the great slogans of the early days of the Nazi revolution was 'Gemeinnutz vor Eigennutz', common wealth before private wealth. It sounds like a statement which science might make, that in judging whether a thing was good for society, one has to consider whether it is good for the whole society and not just for a few individuals. But the Nazis shifted its meaning ever so slightly, and turned it thereby into a totally unscientific notion. What they meant by the common wealth was not the actual good of all the members of society, but the good of some mysterious entity, society as a whole, something which included but transcended the mere individuals. What was good for Germany, the Nazi Commonwealth, might be very bad for the Germans, the common people. They made the group into a person; something which science has always refused to do because it does not make sense.

The Nazis recognised, more clearly than many scientists, that science can only be concerned with the tangible welfare of real people, that the overriding personified state is not a scientific entity. They set themselves to destroy science and build up in its place a pseudo-science which would never overstep the boundaries fixed for it by their political beliefs. They persecuted scientists not only for being Jews, or democrats, socialists or Communists, but for being scientists in the ordinary meaning of the term.

'The new science,' says Dr Bernhard Rust, Minister of Science, Education and Culture,[40] 'is entirely different from the idea of knowledge that found its value in an unchecked effort to reach the truth. The true freedom of science is to be an organ of a nation's living strength and of its historic fate and to present this in obedience to the law of truth.' Notice again how near the second sentence is to being true. Interpret 'nation' not in the Nazi way as an entity itself, but as an organised collection of people; put, instead of 'present', the words 'further and advance', and one has something which science could agree to. It was the strength of the Nazis that they were sufficiently aware of the social problems of today to say something relevant to them; but they invariably said just the wrong thing. Always their

CHAPTER VII

IS COMMUNISM SCIENCE?

Of all the political systems which are alive at the present day, Communism is almost the only one whose influence is mainly due to the force of its ideas and not the mere number of its adherents. It has worked out a consistent body of theory which ties all its beliefs together with the threads of reason. That alone entitles it to careful and sympathetic examination by scientists. And their sympathy must be increased when they notice that Communists profess the greatest admiration for science, and carry this admiration into effect in the Soviet Union, the only country where they have had the power to implement their ideas for more than a few months.

From being an extremely backward country in science, Russia has in twenty years become an extremely important one. In some sciences, particularly the newer ones such as genetics, it is already producing as good work as any country in the world; in older disciplines, in which other countries had already a long start over it, it has not yet been able entirely to catch up. It is often judged on an unfair standard, which neglects the backwardness of the pre-Revolutionary science from which the present development had to start. The Russian achievement should be compared with that of, say, India or South America, rather than of Britain or the United States. The illiteracy from which it has sprung, and the rapidity with which it has been forced forward, is the explanation of the undoubted scandals which have occurred, such as the success (which may now be receding) of the careerist efforts of Lysenko and Prezent, a pair of able charlatans who did much to corrupt the growth of Soviet genetics. But on any reasonable standard its achievement is remarkable. Granted that some of its young men are in too much of a hurry; granted that it lacks the ballast of experienced and critical older men; but there is no possibility of denying that it has come a long way in a short time and that more official encouragement has been given to science in the U.S.S.R. than in any other country of the world in recent years, or indeed at any time in the past.

These very practical deeds are not confined to founding research institutes and financing professional scientists. The government has used its influence to put science and reasoned thought over to the public. I quote from Haldane:[41] 'The intensity of the interest taken in philosophy in the Soviet Union may be gauged by the statement, which I believe to be true, that in 1936 one hundred thousand copies of a translation of certain of Kant's works (I cannot believe they were his complete works!) were printed, and the whole lot sold out.' Ruhemann, describing conditions at about the same time, writes: 'Every Soviet newspaper prints leading articles on scientific and technical subjects and the results of science and engineering are front page news. . . . Well-stocked scientific and technical bookshops are as frequent in Soviet towns as tobacconists are in London.' This, if true, is enough to make any scientist, who is not terrified at being dragged out of his laboratory into the light of day, throw up his hat and shout.

Some of them, and some of the most brilliant, responsible and enterprising, have done so. Bernal finishes his book, *The Social Function of Science*, probably the most important book on this subject to appear in recent times, with the following paragraph:

> Already we have in the practice of science the prototype for all human common action. The task which the scientists have undertaken – the understanding and control of nature and of man himself – is merely the conscious expression of the task of human society. The methods by which this task is attempted, however imperfectly they are realised, are the methods by which humanity is most likely to secure its own future. In its endeavour science is communism. In science men have learnt consciously to subordinate themselves to a common purpose without losing the individuality of their achievements. Each one knows that his work depends on that of his predecessors and colleagues, and that it can only reach its fruition through the work of his successors. In science men collaborate not

because they are forced to by superior authority or because they blindly follow some chosen leader, but because they realise that only in this willing collaboration can each man find his goal. Not orders, but advice, determines action. Each man knows that only by advice, honestly and disinterestedly given, can his work succeed, because such advice expresses as near as may be the inexorable logic of the material world, stubborn fact. Facts cannot be forced to our desires, and freedom comes by admitting this necessity and not pretending to ignore it.

These are things that have been learnt painfully and incompletely in the pursuit of science. Only in the wider tasks of humanity will their full use be found.

This is a very fine statement of the aims and method of science, whether or not one agrees that in acting in this way scientists are behaving like Communists. Other authors[42] have been even more enthusiastic, and put the relationship the other way round, claiming that Communism is the application of the scientific method to political and social affairs. This is a claim which cannot be dismissed out of hand, not only because it has been made by reputable scientists, but also on account of the publicly expressed and official valuation of science in Soviet Russia. It is necessary to discuss sympathetically whether Communist policy is decided by the application of scientific thought and whether that thought, supposing it to be scientific in method, is being correctly applied. I do not, however, intend to enter into detailed controversy about the tactics of the Communist party as a political body in the present situation. The scientific approach, which I am trying to follow, is primarily an intellectual affair concerned with the rational basis on which those tactics are based.

The Communist system of thought, as is well known, was first worked out by Marx and Engels in about the 1860's and was subsequently developed by many authors of whom the most important are those two extremely practical men, Lenin and Stalin. The best recent English discussion of its relation to science is in Haldane's book, *The Marxist Philosophy and the Sciences*. As Haldane points out, Marxism is not a complete and finished system of dogma. 'Marxism,' he says, 'is not complete, not a system, and only in the second place theoretical. It is not complete because it is alive and growing, and above all because it lays no claim to finality. The most that a Marxist can say for Marxism is that it is the best and truest philosophy that could have been produced under the social conditions of the mid-nineteenth century. It is not primarily a system, but a method.' And he quotes Engels: 'The sovereignty of thought is realised in a series of extremely unsovereignly-thinking human beings; the knowledge which has an unconditional claim to truth is realised in a series of relative errors; neither the one nor the other can be fully realised except through an endless eternity of human existence.' This is a claim of exactly the same kind that science makes. It, too, knows that it has never discovered the whole truth, and that each individual can only hope by his efforts to come a little nearer than his predecessors to a full comprehension of the processes he is studying; it, too, is a method of approach, and not a final body of hard-and-fast doctrine.

In its second most important point Marxism is also in perfect agreement with science. It is a materialist philosophy. That does not mean that it believes that everything in nature is a machine in the sense that a motor-car is one, or that it is only the ultimate physical elements, atoms or electrons or whatever they may be, which are of any significance and all the rest is mere froth. It means merely that there is a world of stubborn reality which we can investigate, and which can be changed by our actions, but not by our thoughts alone. As Lenin wrote:[43] 'The sole property of matter – with the recognition of which materialism is vitally connected – is the property of being objective reality, of existing outside our cognition.'

Many scientists would disagree with that statement in theory, and more would argue that it had no meaning and was simply nonsense. But, like everyone else, those who argue that the world is a product of our own minds cannot, by taking thought, add a cubit to their stature; and the people who argue that it is meaningless to ask whether a hare is real or not would agree with Mrs. Beeton that the first step to jugging it is catching it, and that is not done by sitting down and hoping.

The next point in Marxist thought is not merely in agreement with orthodox scientific views, but is, I think, in advance of them, and states clearly and definitely an idea which science is only just beginning to recognise. Everything in the world, this part of Marxism states, is essentially and necessarily changing and developing. The typical thing one must expect to find in nature is not something like a stone, which apparently stays the same for ever, but something like a flame or an animal. As a recent non-Marxist philosopher has put it,[44] 'Our knowledge of nature is an experience of activity'; but forty years before that was written, Engels[45] was already writing of 'the great basic thought that the world is not to be comprehended as a complex of readymade things, but as a complex of processes.'

It is a pity that, for historical reasons, Marx and Engels chose a dialectical argument as the typical example of the process by which things interact with one another. By that is meant an argument in which one man asserts one thing, and his opponent the exact opposite. It was the characteristic method of thought of the Middle Ages, a technique which has been given a thousand-year trial and has produced practically no increase in man's understanding of nature. But this unfortunate choice does not obscure the enormous advance which the Marxists made by their insistence that change is an essential part of the world.

Much of the recent development of science seems to have been towards a view of this kind. As I understand it, the basic ideas of modern physics, quantum mechanics and the theory of relativity, do actually describe the world in terms of processes and not in terms of static things. Certainly in biology, a field which I know more about, the process view (what is called dialectical materialism as opposed to mechanical materialism) is more or less unavoidable. Living things are not mere machines; they are essentially developing and changing things, growing from the egg to infant to adult, and dying, and linked with others in a succession of individuals which show the long-range changes of evolution. These are incontrovertible facts; but I believe biology at present under-estimates their importance, and would be well advised to give them something more like the emphasis which the Marxists urge.

The basic notions of Marxist philosophy are then almost, if not quite, identical with those underlying the scientific approach to nature; there is certainly nothing in them which could cause scientists to reject the rest of the Marxist system out of hand.

I cannot pretend to judge the value of Marxism in other fields in which I am not specially trained with the same assurance as I can in science. But one can make an attempt to apply scientific criteria to its pronouncements on economics and sociology, which are after all subjects about which everyone must have picked up a certain amount of information. In the sphere of economics even the layman can recognise that Marxist theorists foretold the coming of the crisis of 1930 at a time when orthodox economists and social scientists had no idea that it was on the way. That seems very strong evidence that the Marxist theory of economics has a great deal in it, and orthodox economists who reject it have a lot of explaining to do. Until they can show a similar success, the Marxists seem to be winning on the practical test.

It is in the Marxist theory of social action that we come to the first point where I believe that the evidence shows that they have made an important mistake; not so much a mistake of theory as one of emphasis. Marx claimed to be able to show from the study of history that the important steps in the development of human societies were brought about by the actions, not of individual men, but by large groups of people; individual kings and statesmen can only increase or decrease the speed with which the will of the dominant class in society becomes effective. There is nothing much to quarrel with in that view; the only point worth arguing about is just how effective individuals may be in speeding up or slowing down the process.

Next, Marx went on to argue that the groups which are effective in society are classes which differ from one another in their economic position, and that it is their economic status which is the most important factor in determining the direction in which they try to alter society. The class which owns the machines by which wealth can be made will want to keep them; the class which does not own them will want to get them.

It seems to me that science is bound to accept that as the ultimate explanation of the broad outlines of human history. So long as one adopts the view that human actions are caused by anything, the only things there are which can possibly act as causes for them are the material facts. If one rejects that, one rejects the whole causal concept, in relation to psychology, and reduces it to a completely arbitrary subject which has no relation to the rest of nature. That is certainly not true of animals, and it would be completely unscientific to suppose that it is true of man.

But – here comes the 'but' – the relation between the external world, the brute facts of existence, and what a man thinks about them, is not at all simple. Marx, living when the sciences of psychology and sociology were in their infancy, thought that one could short-circuit the complications and make the simple rule that, on the whole, men would see where their material advantage lay, and go after it. It seems to me that the facts one can observe around one show that that is not true.

The number of people who do not own the actual machines by which wealth is manufactured out of raw materials has, as Marx foretold, grown enormously. But they have not been, as he said they would be, forced down to the level of existence of the lowest wage-earning proletariat. And the greater proportion of them do not go after their immediate material interests by trying to seize control of the machines. The only place where this has happened in a big way was Russia, which, at the time it had its revolution, was at about the same, or even at an earlier, stage of economic development as the Western Europe of the 1880's, in which Marx was writing, and sociologically still more backward. In all the other countries the development of the economic system has proceeded much on the lines Marx envisaged, but the psychological effects of this evolution of the non-owning class have been quite different from what he thought. Instead of being primarily concerned to better their economic position, which was not nearly as bad as Marx expected, the majority of them seem in practice to have been much more worried about the emotional and social chaos which capitalism produced, and which was much worse than Marx foresaw. Instead of socialist revolutions, the Fascist movement spread over practically the entire highly industrialised world except for England and America.

Such a mistake, if it is a mistake, would not deprive Communism of the right to be considered a scientific movement. All the sciences have made mistakes; all that it is necessary to do is to correct the mistake and no great harm is done. But unfortunately this is a mistake which it is very difficult to put right. Communists are working as representatives of the great mass of mankind who, as they correctly deduced, would be forced out of possession of the means of making wealth. But they thought that these dispossessed would be put in the position of lowly paid wage-earners, and this conviction is so strong that they consider the wage-earners as the only true working class, the only people who genuinely have lost their control over the means of production. This is not at all true. Very large sections of the middle classes are no more in control of the means of production than are factory hands; they are 'the masses' just as much as are the typical proletarians.

For some reason, perhaps even because of an unconscious realisation that they have made a mistake, Communists have made the service of the working class into the central focus with which their emotions and actions are integrated; and for this purpose they define the working class as the proletariat which Marx foretold. That is the fundamental reason why in my opinion, science cannot admit that Communism is a scientific doctrine. Every approach to the world has its own criterion of value, some crucial test by which in the last analysis it judges whether to accept or reject a statement or an opinion. For science this is the critical experiment; the final test is whether a thing is true when tested in practice. For Communism the final test is something else, service to the interests of the working class. It is not part of Communist policy to make false statements, or to distort evidence, but they are less heinous crimes than to be a traitor to the working class; and this in spite of the fact that at least nine-tenths of even the wage-earners of England do not believe in Communist doctrine.

So long as loyalty to the working class is its final test of value, Communism cannot claim, as it has done, to be the application of the scientific attitude to politics. That would be true even if its theory of the class structure of present-day society was sociologically adequate, and it is even more to the point if its theory is, as I suggest, incorrect. In my view, Communist doctrine falls into three parts. Their scientific philosophy, which I have some competence to judge, seems to me profound, and an advance on anything which has gone before. Their economic theories I have no special qualifications to assess, but there seems to be a case for the orthodox economists to answer. Their sociological and political theories it is the duty of every single citizen to judge, and in my judgment they do not fit the facts or lead to successful action in Western Europe or America.

Up to this point in the dicussion, I have been using the word 'Communist' to mean any one who acknowledges that he consciously bases his political views on the works of Marx and Engels. This is a wider use than that sanctioned by the official Communist party, which includes only those who fully accept the later political thought of Lenin and Stalin. The Stalinist Communists have even less claim to be scientific politicians than have Communists in general, since their views are fundamentally coloured by the particular problems that happen to have arisen in Soviet Russia. Few non-Communists will forget the extraordinary contortions into which they were led by trying to follow the party line at a time when Russia (perhaps sensibly from her point of view) was trying to wash her hands of the war. In point of fact, it is rarely realised in Western Europe how different the Russian culture is from anything we are used to. The Soviet attempt to raise an illiterate and superstitious rural peasantry to the height of a great industrial co-operative nation within a few decades, and in the face of a hostile world, is a magnificent and inspiring effort. But many of the problems they have faced, and still more of the methods by which they have met them, are almost outside our comprehension. To accept every one of their political principles as applicable to Western European or American circumstances, is merely a sign of a lack of sociological imagination. And this is the more so, since it has become clear, for instance in the international debates of U.N.O., that the Russian deals with political controversy much more as an elaborate chess game – much more in the spirit one expects from Egypt or Persia, where a statement is not so much an expression of fact or opinion, but rather a pawn advanced for purposes of manœuvre – than in the somewhat more sincere way we have grown used to.

All this does not necessarily mean that science and Communism must immediately part company. Although an objective view of society would seem to indicate that the Marxist conception of the working class is unsatisfactory, many of the ideals for which the Communists strive in the name of the working class seem to me, as I shall argue in the next chapter, to be the same ideals as the scientific attitude would lead one to formulate. And one can accept their view that support for such proposals is more likely to be found among the non-owning than among the owning classes. The general ideological or cultural outlooks of science and of Communism are, in fact, very close to one another. The reasons why it seems to me impossible to accept Haldane's suggestion that they are identical are based not on general considerations such as those which lead to the rejection of Nazism, but on some of the particular deductions the Communists have made from their fundamental theory, such as their theory of the working class discussed above.

CHAPTER VIII

SCIENCE AND POLITICS

Most people's political opinions are like those bags one meets in statistical problems; full of balls, red balls, yellow balls and true-blue balls, mixed in arbitrary and unknown proportions. The scientific approach must try to treat the problem in a more orderly and systematic way. In this book I can only attempt to describe the way in which a scientific study of such matters would proceed. Its first step would be to try to decide on the aims of society; its next to define the nature of our existing difficulties, not only material but also intellectual. Only then could it begin to make a rational plan of how to unravel the knots into which we have tied our lives.

The characteristics of the scientific attitude have been discussed in different connections all through the earlier part of this book, and there is no need to go through them in detail again. The most fundamental point is that science is concerned to discover how things work, and its test of truth is that it can make them work as it wants to. In order to do this, it is necessary to adopt what at first sight seems an unduly matter-of-fact, cool attitude to many traditionally accepted ideas. Science owes no loyalty to institutions, such as, for instance, an old University or a political party. That does not mean that it must necessarily destroy such loyalties. But instead of regarding them as things valuable in themselves, and sufficient justification, without further argument, for action of some kind, a scientific attitude would regard them as feelings which fulfil certain functions in the whole psychological make-up of the persons concerned. These functions may be very important parts of a valuable attitude, in which case the traditions would be accepted by science as valuable things; but the value does not derive from them themselves, but from the fact that they are part of a mechanism which produces a valuable result.

Science derives its idea of what is a valuable result from its knowledge of the nature and behaviour of things. This sounds an extravagant claim; philosophers have been looking for a definition of value for centuries without finding one. But the scientist's idea of value is much more relative and restricted than the philosopher's. It does not pretend to be 'the Good' from any absolute and final point of view. It is merely the good from the point of view of what is going on here and now. For instance, when biologists discover that animals which were once lively and active crab-like creatures have in the course of time evolved into mere shapeless sacks filled with eggs and sperm, and parasitic on other animals (*Sacculina*), they have no hesitation in speaking of this as a degeneration. By that they do not mean that the parasite is any worse than the free-living animal in the sight of God or in any absolute way; they have no views on such questions. All they can mean is that, judged by the general course of evolution as it is seen in other animals, the parasites have gone backwards and not forwards.

In exactly the same way, observation of the course of man's evolution from the animals, and of the historical development of the human personality, provides a criterion by which we can decide between advance and retreat; it gives us a direction. The most characteristic difference between the behaviour of man and that of other animals is that man can react to much more varied and much slighter differences between things, and particularly to much more subtle differences in the relations between them. A monkey can just succeed in reacting to the relation between a few boxes and a banana dangled out of his reach; he can see the point sufficiently to learn to pile the boxes on top of one another, climb on top, and eat the banana. Man, I think one would agree, can do better than that.

During the evolution of mankind from the earliest Palæolithic savages to the present day, one of the continuous developments has been man's increasing control over his surroundings, attained by the gradual application to more and more things of the orderly matter-of-fact thinking which is science. There may be other such steady progressive movements. If so, it is for someone else to say what they are. The scientist can be content to point out, what few would deny, that the increase of scientific control is one of them, and that therefore the further development of the scientific attitude is a forward movement and not a backward one.

Haldane has said:[46] 'We are part of history ourselves, and we cannot avoid the consequences of being unable to think impartially.' But far from trying to avoid thinking as parts of history, that is much the most important and difficult task which confronts us. It is because they deny this that one can be convinced that the Nazis were fighting against the whole weight of man's evolution.

Haldane quotes a typical example of Nazi thought, from Dr. Johann von Leers: 'After a period of decadence and race obliteration, we are now coming to a period of purification and development which will decide a new epoch in the history of the world. If we look back on the thousands of years behind us we find that we have arrived again near the great and eternal order experienced by our forefathers. World history does not go forward in a straight line, but moves in curves. From the summit of the original Nordic culture of the Stone Age, we have passed through the deep valley of centuries of decadence, only to rise once more to a new height. This height will not be less than the one once abandoned, but greater, and that not only in the external goods of life.'

We see once again that extraordinary and complete vacuum at the base of the Nazi edifice; for we know almost exactly nothing about the culture of the Stone Age, Nordic or not. A man who can find nothing to admire between that time and this rejects the whole of the human development of which we are conscious; and the falsity of doing so is apparent if one looks a little further, since it is just in the deepening of consciousness that man differs from the animals from which he is derived. Man is the only animal to discover the secret of getting results in this material world, which is to let one's actions be governed by an objective analysis of the situation. Our task now is to enable this analysis, which we call science, to get results when it is applied to politics.

But perhaps there is some excuse for the Nazis' muddleheadedness. There has undoubtedly been a far-reaching breakdown of the accepted standards and values on which Western European civilisation has been based. During the war we found a common purpose on a clear, though limited, objective. Now we go back to peace, and face again the conflict of ideals which so nearly reduced us to impotence before. The most definite scientific data on these comes from America. One classic account of them has been given in the two great studies of a typical American town, *Middletown* and *Middletown in Transition*. Their author[47] sums up his findings thus: 'The sense of the augmented too-bigness and out-of-handness of our contemporary world is neither illusion nor merely another expression of this recurrent restlessness of man in civilisation. While unprovable because of our inability to relive intimately the moods of the past, it appears probable that we today are attempting to live in the most disparate and confusing cultural environment faced by any generation of Americans since the beginning of our national life.' That is not true of Americans only, but of Europeans as well. Here are some of the examples he gives of conflicting social beliefs which he finds, by careful investigation, to be particularly important in the lives of Americans; it is easy to recognise them as important in one's own life, too.[48]

'1. Individualism, "the survival of the fittest", is the law of Nature and the secret of America's greatness; and restrictions on individual freedom are un-American and kill initiative.

'*But:* No man should live for himself alone; for people ought to be loyal and stand together and work for common purposes.

'2. Everyone should try to be successful.

'*But:* The kind of person you are is more important than how successful you are.

'3. The family is our basic institution and the sacred core of our national life.

'*But:* Business is our most important institution, and since national welfare depends upon it, other institutions must conform to its needs.

'4. Religion and "the finer things of life" are our ultimate values and the things all of us are really working for.

'*But:* A man owes it to himself and to his family to make as much money as he can.

'5. Life would not be tolerable if we did not believe in progress and know that things are getting better. We should, therefore, welcome new things.

'*But:* The old, tried fundamentals are best; and it is a mistake for busybodies to try to change things too fast or to upset the fundamentals.

'6. Hard work and thrift are signs of character and the way to get ahead.

'*But:* No shrewd person tries to get ahead nowadays by just working hard, and nobody gets rich nowadays by pinching nickels. It is important to know the right people. If you want to make money, you have to look and act like money. Anyway, you only live once.

'7. Honesty is the best policy.

'*But:* Business is business, and a business-man would be a fool if he didn't cover his hand.

'8. Education is a fine thing.

'*But:* It is the practical man who gets things done.'

And so on. They are things that we all believe, sometimes; and they contradict one another. People always have believed contradictory things; probably no one in the history of the world has ever had a completely consistent set of beliefs. But there are grounds for thinking that the contradictions have never appeared so obvious, have never been so much in the forefront of consciousness and caused such mental strain as they do today.

The evidence suggests then that the need for a coherent and generally accepted set of beliefs is one which will have to be met by any satisfactory solution of the political problem. Science starts at a great advantage in meeting this need, since many thousands of men have laboured for several centuries to make it self-consistent.

Students of primitive cultures, investigating the social mechanism of savage tribes and the way in which they are organised, have separated two main types of beliefs, both of which have to function smoothly and without contradictions if the society is to be stable. They[49] have called them the eidos and the ethos (Greek scholars excuse!) The eidos is the name for the whole set of theoretical beliefs by which the organisation of the tribe is justified. For instance, that the god who created the tribe emerged from a certain hole in the ground, and then mated first with a crocodile in the river and then with a bird in the forest; and the descendants of the first marriage belong to the crocodile totem, and those of the second to the bird totem; and the crocodiles are fishermen while the birds gather fruits; or some such rigmarole. The ethos is the system of emotional attitudes which are socially recognised as proper in different circumstances; for instance, that mothers-in-law are to be treated as a jest in the music-hall, but taken seriously at home. For our society, of course, also requires an ethos and an eidos, and much more complex ones than are found in small out-of-the-way tribes.

Science has a perfectly good eidos to offer. In fact, it is its duty to provide a set of reasons for all natural phenomena, including our own behaviour. Its theoretical structure has the advantage that it is founded on a careful investigation of the world, and will not, therefore, be easy to catch out in contradictions which might upset the smooth running of a society based on it. It has also a definite ethos, as anyone will know who has heard scientists discussing with one another. It is an ethos which allows plenty of scope for individuality. In fact it encourages people to put forward their own points of view; but it insists that they should support them by reasons which other people can verify, and that they should be willing to accept the judgment of critical experiments as to whether they have made out their case. *It is an ethos based on the recognition that one belongs to a community, but a community which requires that one should do one's damnedest to pick holes in its beliefs. I know of no other resolution of the contradiction between freedom and order which is so successful in retaining the full values of both.*

The ethos and eidos of science are things which all scientists are in substantial agreement about. There is no such agreement about the scientific solution of the economic problem, although everyone would admit that in recent years our economic machinery, particularly the machinery of distribution, has not been working as it ought. The crucial point about which controversy rages is whether this is due to the fact that in our society the incentive to production is private profit and not, at least directly, the needs of the community. Now the scientific approach definitely defines the criterion on which the merits of the two sides are to be judged. For the sociologist looking at a society, the thing which is good for the society is simply what is good for the people of the society; there is no more to it. There is no question of justifying a society by saying, as was said until a few generations ago, that it was so ordained by God, and that people should be satisfied in the station to which God has called them. Such an argument is foreign to the scientific attitude. A scientist must agree that production with a society should be production for the society as a whole. The important question for science is whether private enterprise is the most efficient method of production for the good of society as a whole.

It is probable that it has been so at some times in the past, in some places. But it was certainly not maximally efficient in Western Europe and America during the last few years before the war, and it has been unable to bring to colonial peoples, such as the Indians, the riches which the full application of modern technique could win from their countries. I think, though not by any means all scientists would agree with me, that the socialists have made out a strong case that they would be able to organise a better system. It is difficult to overlook the considerable increase in wealth in the Soviet Union, the only socialist country, throughout the twenty years during which the capitalist countries suffered an appalling depression and finished up in a war. As soon as some governmental control (i.e. socialism) was imposed on American industry in the early years of the war, its productivity leapt up at a phenomenal rate. In general the whole of evolution is concerned with the gradual increase in conscious rational control over ever more complex fields of behaviour. It seems inevitable that at some time the economic forces in society will have to be organised by human thought instead of by the automatic 'laws' of supply and demand. This is in fact admitted in practice by the capitalists themselves, who have been doing their best for many years to dicker with their system by means of tariffs, quotas, price rings, etc. Their lack of success up to the present suggests that someone else should take over the job.

To talk of economic organisation, as is often done, in terms of revolution is beside the point. A revolution is a method of redistributing power, not of organising production, as one can see in the Soviet Union, which went on developing its economic machinery long after the actual revolution had been won. One can never expect to see more than one or two steps ahead in the direction of a satisfactory plan.

No one who appreciates the problems of present-day Europe from a scientific angle can be very impressed by any solution which is primarily in political terms. The distinctions between the usual political parties – Conservative, Labour, Radical, Communist and the rest – are not, I think, of crucial importance to the scientific mind. The important line in politics is between those who judge the value of a society by its efficiency in maintaining itself and by its advance along the whole line of human evolution, and those who judge it by some other criterion, whether based on mysticism, nostalgia for the past, or motives of personal advantage.

From the point of view of science, this is the first great cleavage. All those who hold the first belief are within reach of reason and intelligent argument, while those who come into the second category can only be touched by propaganda addressed to their emotions. It is not a distinction which can be easily and automatically employed. Most people whose beliefs have an emotional basis will aver that they can produce convincing rational grounds for them; and men whose opinions are suspect because they seem founded on a trust in emotionally potent forces, such as private property, may really be expressing a rationally derived but unorthodox analysis of society. But the difficulty of using the criterion does not destroy its importance; important things are rarely easy.

The cleavage I have mentioned cuts right across the normal political parties; even, I think, across such a homogeneous body as the Communists. But it is true that one is likely to find a greater proportion of non-property owners than of capitalists on the same side as science, since they will not be pushed in the other direction by motives of personal gain.

It is not likely or desirable that scientists will, as a body, set themselves the task of forming a political party. Science can only be effective as a set of ideas which permeate the public mind. The endeavour of scientists who wish to see their ideas applied in practice must be to encourage and speed up this permeation. The task is primarily one of propaganda.

Its magnitude is formidable. Only a very small number of people have a clear idea of what a scientific argument is like, and what sort of criteria a scientist uses to decide whether a thing is true enough to act on. I have in front of me a memorandum on the popular conception of the scientist. It was drawn up during the war for a group of us who were interested in the matter, and although it is not based on a comprehensive survey, it is the work of a man with considerable experience in studying the popular mind; I quote from it by his permission.[50]

'The following examples, taken at random, give a small picture of the huge barrage which, from boyhood magazines and the Wizard of Oz onwards, builds up the layman's picture of scientific method and purpose. . . . In *John Bull*, September 28th, there is an article typical of the treatment in the popular weeklies, which reach a majority of homes (about 10 per cent of the whole population are regular readers of *John Bull*). It is headed "THAT MAGIC SWITCH", and starts:

> "People still find it almost impossible to believe that they are in possession of anything they cannot see. The Power of the Mind leaves them sceptical.
>
> "Any doctor will confirm that mind is master of the body. *It is now a settled principle of science.*"

'This article goes on to give a particular anecdote about a scientist. The sort of anecdote which, repeated a million times over, has produced an image of the scientist as someone removed from ordinary human realities, an almost supernatural figure:

> "Once, when Sir Isaac Newton was giving a dinner, he left the table to get some wine. On his way from the cellar he became lost in reflection on some philosophic problem, forgot his errand and company, and was soon hard at work in his study."

'Nearly every boys' magazine is crowded with scientific figures; the only boys' magazine which refused to deal with Martian problems has recently folded up. Usually the scientist is a brilliant, infallible, but subsidiary hero; the virile young millionaire-explorer is boss. . . . Examples of the above sort of thing could be multiplied indefinitely. So could examples of the super-invention, transcending the earth, which is by no means confined to boys' magazines – e.g. the weekly with the biggest circulation of all, *Everybody's*, has as its popular strip cartoon hero Buck Rogers, the boy with the space-yacht, who defies the Moon-Priests, the Island of Doom Martians, captures girl-friend Alura from another planet, etc. . . (speaking of a best-selling novel). Throughout the story the magnetic eyes and master-mind of the scientist operate, but, *as is nearly always the case* in such stories, the scientist never emerges with any definite character, normal activities, hobbies, political views or love life. . . . This exploitation (in advertisements) of the word "scientific" to give prestige all helps to increase the magic of science, "something difficult to understand but sure to be correct". And this often actually hinders any propaganda for science by making laymen believe, but also believe that they cannot understand how the marvellously clever scientists did it. For instance, in circulating the synopsis for a recent important series of broadcast talks for adult discussion groups, on the Institutions of this Democracy, the B.B.C. gave a full explanation of all the talks except for the one talk about science, where the only comment was a note in brackets saying that this talk would be given if the subject could possibly be explained in terms which were "not too technical". To sum up: In general there is an enormous popular misconception about science, coupled with a considerable unorganised admiration for science as a whole and a semi-humorous, semi-reverent attitude to scientists as people.'

There is not much basis there for implementing science's contribution to political life. And the artists and writers who have been working towards what I have called the scientific attitude are in no better position to throw their weight about; the general public, I think, though I have no special data on the point, consider them just as mad as scientists and rather less useful.

But the situation is not nearly as hopeless as one might think from that report. In the first place, the social effect of a cultural movement, such as science, is very largely produced unconsciously. People who think of scientists as wizards with magnetic eyes may still, unconsciously, show an influence of scientific thought when they come to decide practical questions. Science is, after all, largely common sense; and a common-sense, matter-of-fact thinking is undoubtedly applied to a much wider range of phenomena nowadays than it used to be. Not, of course, that the range is nearly as wide as one would like. But at least the basis is there. When people discover what the method of science is, they find it is something they have known all along. They have not got the same resistance against it as they would against some totally new kind of thinking, such as Yoga.

The most effective way of convincing people of the value and usefulness of science will be the performance by scientists of a few useful and convincing pieces of analysis of social affairs. Within some fields, scientists have already shown that their particular methods of handling practical problems can become an important addition to the older techniques. You may be tempted to ask whether, when we come down to brass tacks, to actual details, the scientist's method of treating the difficulties of society would differ from anyone else's. I should answer that, in theory, it scarcely does differ from the way most practical men allege that they set about a problem; but that in practice it turns out to be totally different both in the steps taken to find out the answer, and very often in the answer arrived at. During the war, Britain made very important advances in the technique of organising the application of science to practical affairs. Probably the most systematic and important use of the new method was in the formation of operational research sections attached to most of the important Commands in the Services. It is worth taking a glance at their functions and achievements, in order to see what kind of thing science can do when it comes outside the laboratory, and in what manner its results may differ from those currently accepted.

Wartime operational research had two roots. One grew up within the military machine. Some of the new devices, particularly radar, were so complicated in operation that the scientists were called from their research and development laboratories to help work out the best methods of using the apparatus in the field. It almost immediately transpired, in the very first days of the war, or just before, that the new equipment also made possible entirely new types of military operation; the whole organisation for countering a bomber raid on this country turned crucially on how quickly the raiders could be detected and the information on their course interpreted in such a way as to make it possible to intercept them. Thus the scientists, as the only people who thoroughly understood the technical possibilities, found themselves more and more forced into the consideration of matters which would previously have been thought to belong entirely to the trained military officer.

The other source of operational research was a spontaneous conviction, among a group of non-military scientists, that the conduct of military operations was bound to throw up many problems for which the scientific method of quantitative and empirical thinking would prove very valuable. This was, in fact, nothing more or less than a demand for the application of the scientific attitude as I have been discussing it in this book. It was first openly voiced in a Penguin Special, *Science in War*, which came out in 1940. Fairly soon, several members of this group had got themselves accepted as scientists attached to various fighting Commands, and, joining up with the most operationally minded radar men, they gradually built up the operational research sections.

The terms of reference of these sections were practically unlimited. At first nobody, not even themselves, knew quite what they were there for. They could do just about what they liked, provided they didn't waste anyone's time. By the time the war got properly 'organised', say from 1943 onwards, there were plenty of odd corners where one could hide away a little unit of staff with comparative safety from dismissal. But these cushy spots were not, in general, on the personal staffs of the chiefs of fighting Commands; and they were by no means so common in the dark days of '40 and '41. It was during that testing time of strenuous effort that the scientists – marked men, in their civilian clothes, among the professionals of a Command staff – had to produce definite results which justified their anomalous and quizzically inspected position. They did so. As the war went on, far from being sacked, their numbers increased; the Americans copied the organisation; operational research sections were set up in other Commands than those which had first been infiltrated; and the scope and responsibility of their work within each Command increased. Science proved a useful helper in the war-making team.

There was, perhaps, no essentially new element in operational research; there were many scattered cases before the war of scientists being called on to act as practical men of affairs.

The novelty in the war set-up was not in the essential nature of the work, but in its scope, in the systematic and organised turning loose of scientific method on to anything which the scientists themselves felt they could handle, and much which, at first, they were scared stiff of touching. The head of an operational research section was the official scientific adviser to the Commander in Chief, with direct access to him; and he usually attended all the Commander in Chief's staff meetings. Thus he knew exactly what problems were troubling the men responsible for taking executive action. His job was to try to help, in any way he could, in arriving at the correct decisions. Occasionally, as in the instance of radar which was mentioned above, arriving at the correct decision involved taking account of complicated technical data. But this was not always, in fact not usually, the case. Scientists turned out to be equally essential for a just consideration of quite non-technical matters, such as the relative merits of searching for submarines in the Bay of Biscay as an alternative to guarding convoys. Everyone knew that there was more chance of spotting a U-boat near a convoy (once an attack had started) than of finding one crossing the Bay. But before one can decide how to allocate one's air fleet to the various bases, one has to know *how many* U-boats are seen per hour of flying in the two regions, *how much* protection can be given a convoy in practice, *what percentage* of submarines will be sunk out of those spotted in the two areas, *how often* the weather will interfere with flying from the two bases, and a whole host of similar questions. The 'science' necessary to answer most of these questions was extremely simple – merely the combing of the records, and taking averages of the past results – occasionally a very little simple algebra, or one of the conventional statistical formulæ as a little pansy trimming. Anyone, with the time, might have done it; when it was done, it had to be straight commonsense or we should never have got the Service staff to swallow it. But after all, much the same can be said of practically all science ever has done – some appears commonsense immediately it is finished, some takes a few years before it is assimilated. The actual doing of it is not much the easier for that! And the important fact for our present purposes is that, when trained scientists did not make operational analyses of this kind, usually nobody did – or if somebody tried to he either didn't know when to stop going into unnecessary detail or left some vital factor completely out of account. To do the thing continuously, responsibly, over the whole field of a Command's activities, and up to time, a team of trained research scientists was indispensable.

Military operational research still provides the best example to date of an intensive and fairly large-scale attempt to use scientific method, not merely to solve technical problems, but to help in defining and solving problems of policy. In spite of the specialised field in which it operated, the tradition and experience which operational research hammered out for itself are full of lessons for future times of peace. One of the most important of these – although it should scarcely be necessary to emphasise it – is that scientists can take part in forming policy as members of a team. Any suggestion that scientists should be allowed to use their training in fields wider than those of sheer technology, is only too often countered with the silly cry that they are demanding dictatorial powers. The expert should be on tap, but not on top, squeal some of those who are on the top now. This is primarily a reaction of fear. Scientists could only acquire undue power in the handling of public affairs in either of two conditions; if the scientific method were so overwhelmingly more efficient than other modes of thought as to put them completely in the shade, or if the non-scientists failed, from personal incapacity, to make their presence felt. Now I do believe that the scientific technique of handling events is the most effective yet invented by man, and that in consequence men trained in these methods will come to play a very large, and perhaps predominant, role in affairs. Even after only one or two years' experience during the war, Air Chief Marshal Sir Philip Joubert wrote: 'One of the problems of a Commander in Chief who had an operational research section was to prevent the scientists taking over complete control of the operation of his squadrons.' But the airman is really paying too fulsome a compliment to 'those University types in flannel bags', as one of them called us. Not even such an advocate of the scientific attitude as I would maintain that other methods of approach have nothing to offer; and in fact the scientists never wished to take, and never came within miles of taking, control of operations. They were, knew themselves to be, and intended to remain, members of a team; with a special skill – important but not all-important – to contribute.

One of the most far-reaching characteristics of the scientific approach to a practical problem is an insistence on deciding exactly what one is trying to do. It is surprising how hard it often is to decide this. To take a fairly simple case. The Air Force convention was that the aircraft in a squadron were being properly and satisfactorily maintained if, on any day, a certain standard proportion were fit to fly. But why that particular proportion – 60% or 70% or whatever the chosen figure? A number of alternative answers have to be considered. It might be that our main concern really was to ensure that there would always be enough aircraft at readiness to meet a sudden emergency. Or it might be that supplies of aircraft were short, and that particular rate of use and repair had been found the most economical in preserving them for a long life. Or again, manpower might be the crucial shortage, and the rate of maintenance be decided so as to use the available labour force as economically as possible. Until one has decided which of these alternatives is the case, one cannot start to rationalise the system. In the actual circumstances of the war, the third alternative – a tight manpower supply – was the most important factor, and the operational research sections were able to calculate new rates of maintenance adjusted to the rates of flying in different squadrons so as to produce the maximum number of hours in the air for the available number of man-hours worked on the ground. The effect was appreciable – an increase of well over 50% at the expense of no more than a little precise thought devoted to asking what one was trying to do.

In how many practical affairs of the day is this essential enquiry omitted? I will not attempt to give examples from the working of business, but I suspect that anyone who knew the facts could easily find them. But in large public questions, examples of un-thought-out objectives are legion. What, for instance, is Britain really trying to get out of the export drive? In general terms, no doubt, a higher standard of life. We export a lot of things to pay for importing a lot of other things; of the things we import, a certain number are worked on and re-exported with an added value; but there remains some residue, of wheat and bacon, pineapples, timber, steel and so on, which we actually use to raise our own standard of living. It is this residue that the whole business is in aid of. But it is extraordinarily difficult to get anyone to say just what quantities of what commodities it consists of (apart from the food items). Or consider an example on a somewhat smaller scale. What are Universities really training their students for? Are they giving a general education in culture and citizenship, or are they trying to turn out technical experts? If the former, what are the crucial lessons which can be taught in the different faculties and how are the courses related to getting these lessons across; or if the latter, what kind of experts, what proportions of the various types of specialists, and so on?

Such questions are admittedly difficult to answer. But it is usually possible to get somewhere near a satisfactory solution, and it is always worth trying. The method of approaching them must be twofold. On the one hand, there are certain very general considerations of value and ultimate objective; for instance, the relative importance of imported material amenities as compared with the spiritual benefits of leisure, or of technological skill compared to a broad humanistic culture. Although such objects are not, in my opinion, entirely outside the field of scientific theory, it is in practice the non-scientists who can usually make the most important contribution regardin them. On the other hand, one must study what is actually being achieved as things are; and this is the main contribution of the scientist. Often it will turn out that our institutions are in practice producing effects which were not at all considered to be their main function. We may think of the Universities as the nurseries of culture, only to discover that they produce very few creative artists and writers, but are the seedbed from which our whole technological achievement grows. Only by combining a criticism in broad philosophical terms with a detailed assessment of the facts can one hope to reach a sensible and practical formulation of the direction which development should take.

It is from this same careful, quantitative, and laborious examination of the facts that one obtains the other two essential bases of an operational research analysis. These are two closely connected data: a survey of the resources available for use, and a description of how they are being used at present. It is only after a decision has been reached as to the end aimed at, and after reasonably complete knowledge has been accumulated about the means on hand, that the most intellectually interesting part of operational research begins. The most valuable method of analysis is usually to try to estimate the effects of introducing variations into the current practice. A complex practical operation always involves a large number of interrelated steps. The scientific method of analysis is to try to find out the relative importance of these in determining the overall rate of the whole sequence. It often demands a good deal of ingenuity to think how data can be found which will enable one to deduce the effect of varying a process at stage B, so that it can be compared with the results of a suggested improvement at stage D, say. But when quantitative methods of estimating such effects can be discovered, they make it possible to decide which are the relative bottlenecks, and where efforts at improvement should be concentrated.

All these principles, when stated perfectly generally, appear the tritest of commonsense. But in practice they would be revolutionary. Consider the problems of India, the Colonies or Palestine. One very large element in all those situations is the need to raise the standard of life of the local inhabitants; yet in even the most serious public discussions, this aspect of the matter is almost entirely elbowed out of sight by an overwhelming concentration on narrowly political considerations. Where can one find a clear statement even of the material objectives of Colonial development? What is it proposed eventually to do with tropical rain-forest, or the inland savannah of West Africa? How often is the future of these countries looked at against a background of a reasonably complete survey of their resources and a technological assessment of the possibilities? Yet the problem of making the tropics fit for civilised life is the largest and most exciting technological problem facing our generation – one which we certainly possess most, if not all, the knowledge necessary to solve. If it were faced with energy, and, as a first step, an adequate scientific staff were given the task of studying the material factors involved, there is no doubt that the political tensions, no longer shouting their war-cries against a background of technological silence, would appear shrunken to more manageable proportions. The lessons of the war experience are in fact just filtering through into this field. A considerable beginning was made during the operation of the Middle East Supply Centre, and the appointment of a Scientific Adviser to the Conference of East African Governors is very much a step in the right direction; but rather a small one compared with the size and interest of the issues involved.

If everything goes according to the book, one should of course expect a great increase in the application of scientific method to the economic problems of England. We have a Government pledged to plan. Perhaps it is being rather querulous, but it is difficult to dispel the last traces of doubt whether the officials who have run the governmental machine for so long really know how to set about the task. The Civil Service, which must bear the brunt of examining the facts and executing the policy, was not organised for the putting into operation of major constructive efforts. Its conventional machinery is far less fitted for such an enterprise than that of the Services, which, as one finds when they get down to the bedrock job of war, are organised to concentrate on getting something done, and have special groups assigned the responsibility of deciding what to do. Moreover the currently accepted expert on planning is the economist, a man whose training is shot through and through with the assumptions of a system in which money is the ultimate criterion and goods are thought of as the mere resultants of the automatic workings of the economic machine after the controls have been set by monetary policy. A scientific approach to planning needs a reversal of this attitude; goods are for it the essentials, and money merely a means.

These fears may be groundless; but it is not obvious that the Civil Servants who plan nationalised industries (or the entrepreneurs who ran them before) are any more above the need of scientific assistance than were the staff officers whose profession was to organise the operation of aircraft for war.

There is no reason why scientific methods should not prove helpful even in the grand political sphere of foreign affairs. Little concerted effort has yet been made to apply them to such matters, but in order to give a concrete example of one, at least, of the ways in which the scientist would try to tackle them, I shall quote at some length from an article by the Anglo-American anthropologist, Bateson, in the *Bulletin of the*

Atomic Scientists for 1946. He is using what the operational research workers used to call 'the model method'; the method, that is, of defining a mathematical system which, it is hoped, will serve as a model similar to the state of affairs under investigation, and from which the probable changes in the real situation could be deduced. Bateson sets out to discuss the progress of an armaments race between nations, to see if we can reach any deeper understanding of the phenomenon which will help us to control it. His mathematical model was actually formulated by L. F. Richardson in a paper published in 1939;[51] we will leave out the algebra, and give the discussion in Bateson's words:

Broadly the phenomena may be summarised as regenerative (or 'vicious') circles of cause and effect, such that A's actions become stimuli for B's actions of the same type, and these in turn become stimuli for further actions by A, and so on. Stated in these very broad terms, clearly every such system might progress at greater and greater speed until it topples over into the state of war. There are, however, a number of other considerations which have to be taken into account:

1. The relation between stimulus and action is complex and is subject to reversals of sign. In Richardson's basic equations it is assumed that A's rate of armament will be proportional to the amount of armament possessed by B. This basic equation he elaborates, examining also the implications of a system in which the effective stimulus would be, not B's total armament but that armament which B has in excess of A. The stimulus factor in the equation is then (B – A), and the rate ofA's armament equals (B – A) multiplied by a constant which Richardson calls 'the defence coefficient'. Such an equation may satisfy these cases in which each side argues 'The other side is getting ahead of us. We must therefore hurry up.'

This, however, is not the only type of argument which stimulates armament. There is also an argument in which the sign is reversed. 'The others are falling behind. Let's get ready to beat them up.' In this case, and also in 'The others are so far ahead of us that we had better appease them' we see a reversal of sign so that the stimulus factor which will make A arm more rapidly is now, not (B – A) but (A – B), and when this latter term is negative in value, we may even see 'negative aggression' in the form of appeasement.

2. It is known that the various cultures of the world differ enormously in the degree to which they are characterised by predominance of one or other of these opposite types of relation between stimulus and aggressive activity. The 'bully-coward' motivation is one which is rather sharply disapproved in American and English life, while motivation in terms of the excess of strength held by the opponent is strongly approved in terms of 'fair play'. In Germany, on the other hand, it is expected as a matter of course that the stronger will take advantage of his extra strength and that

the weaker will submit. In Russia, preliminary studies indicate that very high value is set upon achieving one's own full strength. It is not the fact that one can beat somebody else that is important but the assurance that one is exerting one's full strength. Conflict – especially conflict against an enemy who is conceived to be stronger than oneself – may help one to this assurance but conflict is by no means the only way of achieving it.

There is thus a vast field, here, for more exact research.

3. We must also consider the motivation which each side attributes to the other. Even though each side may be actually motivated in terms of the other's apparent extra strength and would actually slow down if they thought the others were weaker than they, each will attribute to the other the opposite type of motive. Each will say 'If we don't catch up, the others will take advantage of our weakness and will say to themselves "let's get ready to beat them up".'

4. All the statements in 1 and 2 above require to be corrected for the fact that the stimulus term in the equations is not the real strength but the apparent strength – the strength of the others as it is perceived or imagined. This figure is subject to two main types of distortion: (a) Increase or decrease due to the actual falsification of the reports and rumours which each arming nation will give to the outside world; (b) Increase or decrease due either to persecutional fears or to unrealistic optimism, and these psychological factors certainly depend upon the cultural conditioning of the individuals concerned, and upon their realistic sureness of their own personal strength. Such a recurrent myth as that of encirclement is no recent development in Germany but is a deepseated characteristic of German thinking. For example in German fictional films we find, instead of the 'chase' so loved by Hollywood producers and American audiences, slow encirclements of the hero. The human imagination can not only exaggerate or minimise the strength of a potential enemy, it can also distort the picture of how that strength would be used.

5. The matter is further complicated by peculiarities of the new weapons. The atomic bomb is not only a saturation weapon in the sense that when it is used the entire defence machinery of the attacked spot is dislocated. It is also saturating in the sense that a given nation need only possess a limited number – a few thousands perhaps – of this weapon to achieve 'complete' aggressive strength. The making of further thousands will not further increase its aggressive power. On the defence side, however, the picture is very different. Even adequacy of defensive preparation is probably impossible. Therefore it appears that we must expect a world in which several powers are saturated with the weapons of attack, but are still making frantic efforts to achieve some degree of efficiency in defence.

What will be the psychological implications of this emphasis upon defence? An increasing fear of being attacked? An increasing belief that attack is the only defence? Or an increasing realisation that war has become an intolerable business? And will every people react alike to this peculiar state of affairs? To these questions, the anthropologist can hazard an answer only to the last, and to this his answer will be 'NO'. It is most unlikely that the nations will react alike. There will be profound differences between them and these differences will be related inter alia to each people's special habit of response to contests involving strength and weakness, attack and defence. Just how these differences will be expressed we cannot say without further research, but we can warn that the nations will certainly misinterpret each other's behaviour and that it is unlikely that these misunderstandings will be of such a kind as to promote goodwill between them.

6. In addition to the factors which make for increased rate of armament among the nations, there is, as Richardson points out, one important factor which tends to diminish this rate, namely the expenditure which each nation must make to keep up the pace. His equations therefore contain a negative term in which the total strength of the arming nation is multiplied by a 'fatigue and expense coefficient'. Supposing these equations to be substantially correct at least for moderate disturbances – Richardson has analysed a series of armaments races and demonstrated surprising regularity – we may then follow him in assuming that whether the system will move toward a steady state or toward infinite armament, will depend upon a rather simple relation between the constants involved. If we adopt the simpler form of the equations and consider the case of two nations, equilibrium will occur if the product of their 'fatigue coefficients' is greater than the product of their 'defence coefficients'.

In terms of this analysis, the result of the atomic discoveries and other great advances in the machinery of destruction can be stated very simply. It has been to reduce the 'fatigue coefficients'. A nation can now become infinitely prepared to destroy its neighbours (though not prepared to defend itself) for the modest sum of less than five billion dollars. The likelihood that the system will reach equilibirum in Richardson's sense is therefore very much reduced, though it is conceivable that the nations, having achieved infinite aggressive armament, might stand glaring at each other for an indefinite period without actual war. But the fact that the techniques of rapid attack have so far outstripped those of defence make us doubt whether such a balanced position could be stable. In such a case, it would be too easy for any nation to succumb to the argument, 'If we attack, the enemy will immediately become weak. Let us therefore beat them up', and too easy also, for each side to fear that the other may be on the point of arguing in this way.

This discussion of various types of operational research will perhaps give some hint of the way in which scientific method could be applied to the major problems of society. There are many difficulties to be overcome before such work is undertaken, and its results put into practice, on a scale commensurate with our needs. Not least is the difficulty of persuading sufficient mature and first-class scientists to leave their laboratories and launch out into this uncharted sea. But not until the attempt is made, on a large scale, can we say that we are dealing with the social life of man (which is most of his life) in a fully responsible way.

CHAPTER IX

LIVING IN A SCIENTIFIC WORLD

What would the world be like to live in if the scientific attitude did become generally accepted as the directing force of our culture? Most people, we are told, think of it in terms of Buck Rogers and his space-yacht, and even the more sophisticated will have their imaginations full of Kipling's *Aerial Board of Control*, or Huxley's *Brave New World*, or some of Wells' more aseptic novels. They are likely to think that even if science could give the Open Sesame to all the riches of Ali Baba's cave, it wouldn't be worth much if the only furnishings were dentists' chairs, complete with drills. It seems as if there would be no place, in the chemically purified scientific world, for the ordinary little enjoyments of life, playing bridge or darts, messing about with a bonfire in the garden.

These fears are two hundred years out of date. In science's early days it could only handle the most simple things; it had to reduce an action like standing a fellow a drink to a matter of giving him a certain quantity of diluted ethyl alcohol of a certain strength. But those days are long past. Science realises now that every detail of human life is extremely complicated; it is connected with every other detail by innumerable threads of habit and custom, of economic interest and emotional feeling. A way of life is a network, full of cross connections. It is worth giving an example of how sociologists try to analyse social behaviour; it will show, I think, that they do not disregard all the colour which gives sparkle and interest to life. In fact, they see so many things involved in even the simplest actions that it is at first difficult to follow what they are driving at. One has to try to look at one's own behaviour from outside, as they have been able to look at the behaviour of men in savage tribes.

Consider, then, going out with your wife in the evening to friends, perhaps to play bridge. According to Bateson, the factors involved are (I will give the technical names):

1. *Structural* – Concerned with how the society is built up; for instance, that wives are equals with their husbands, and accompany them to social functions on equal terms, instead of staying at home, or coming to act as servants.

2. *Eidological* – Concerned with the way people think in that society; for instance, they think of their acquaintances as human beings like themselves, with faults and virtues, desires and feelings and so on, but not as potential witches to be on one's guard against, or as mere ciphers in an elaborate ritual of politeness.

3. *Affective* – Concerned with the emotions and feelings involved in the meeting, the friendship for the people visited, the gaiety or sobriety of the party.

4. *Ethological* – Concerned with the general system of feelings which society considers appropriate; that one behaves politely to people one doesn't know too well, but may take more liberties with intimate friends; that a certain amount of assertiveness and drive is expected from men, but that women should be more homely and sympathetic.

5. *Sociological* – Concerned with the stability and integration of society; friendly visits bind people together with bonds of common interests and affections.

6. *Economic* – Perhaps you are taking your wife to call on the boss, and the impression you make will affect your chances of promotion.

7. *Developmental Psychological* – Concerned with the education which has brought you up to be a person who likes to go out for casual enjoyment, instead of someone tormented by conscience, which might drive you to spend the evening in prayer, or someone with a passion for improving your mind, so that you would rather stay at home with a course from a correspondence college.

No one would pretend that those were all the factors involved in your bridge party, but they are enough to start with. Science does not consider the whole thing simply as changing the space-time co-ordinates of a few bits of pasteboard. It acknowledges that it is a part of life, and must be handled with the gentleness which that makes necessary. It is the present ferocious monkey-house civilisation, brilliantly scientific in spots and plain brute in others, which really threatens the quiet and comfortable life, even in times of so-called peace. As an American poet has written:[52]

> They are able, with science, to measure the millionth of a millionth of an electron volt.
>
> THE TWENTIETH CENTURY COMES BUT ONCE
>
> the natives can take to caves in the hills, said the British M.P., when we bomb their huts.
>
> THE TWENTIETH CENTURY COMES BUT ONCE
>
> . . . . .
>
> the lynching was televised, we saw the whole thing from beginning to end, we heard the screams and the crackle of flames in a soundproof room.
>
> THE TWENTIETH CENTURY COMES BUT ONCE
>
> You are born but once;
> you have your chance to live but once;
> you go mad and put a bullet through your head but once.
>
> THE TWENTIETH CENTURY COMES BUT ONCE
>
> Once too soon and a little too fast;
> once too late and a little too slow, just once too often.[5]

That's not what we want to go back to now the war is over.

But however much one may point out that a world imbued with a scientific outlook would be tolerant, sympathetic and sincere, there will remain many people who will regard the whole thing with horror. The old ways in which one was brought up have an extraordinary hold, and for many people any idea of change is hateful. It is resented with an intensity altogether out of proportion to the actual effect, as far as the outsider can judge, which it is likely to produce on ordinary lives.

A society, in fact, does not behave entirely like a rational human being. As Ruth Benedict[53] says, 'Tradition is as neurotic as any patient; its overgrown fear of deviation from its fortuitous standards conforms to all the usual definitions of the psychopathic.' So it is probably stupid to expect any new way of life, whatever its advantages over the old, to be thankfully accepted even by all the people to whom it will most obviously do good. Trying to cure the ills of society is rather like dosing a cow; you want one man to hold its mouth open while another blows the pill to the back of its throat down a cardboard tube; and it has the same danger, that the cow may blow first.

Advocates of a new order have to expect more or less violent opposition. They may console themselves, if they have confidence in their remedy, by a fact revealed by the psychoanalysts, and, I think, confirmed by common sense. People only object really strongly and irrationally to something if it touches a part of their personality which is out of order, which they would themselves object to if they could dare, or could make the effort, to face it. Their 'instinctive' rage at the new proposition is so intense only because they really know there is something in it, but do not want to let themselves see it.

Of course, the same thing applies the other way round – to the advocates of the new conception. They become preachers; people who make too much fuss and shout too loud. The louder they shout, the more coldly the others turn them down; and the more indifferent and cloddish, or the more rancorous and spiteful, the rest of the world seems, the shriller they yell. An unlovely but ordinary process, for which sociologists have coined the unlovely but out-of-the-way phrase, complementary schismogenesis;[54] something difficult to cope with, both in pronunciation and actuality.

The character of the scientific world, as a place for the ordinary man to live in, will depend enormously on how far this process of mutual blackguarding has gone before the new ideas become firmly accepted. If it is comparatively mild, the world will be a fairly easy-going and tolerant place; if it becomes virulent, the scientific attitude will be forced into a crusading, revolutionary ardour which is foreign to it, and which will make it a much less equable climate of opinion for those who do not share its enthusiasm.

There is no reason, *of an intellectual and cultural kind*, why this should happen in England or America. In both places the scientific spirit is already a powerful influence, and there is no other alternative outlook on the world which seems strong enough to cause much friction; science's battle with the only other strongly entrenched system of thought, that is to say, religion, is now ancient history.

But, of course, the snag is that other things are not equal. A scientific reorientation of society would involve economic changes, and anything of that kind is sure to be resisted by most of the people who are adversely affected by it. At the present time it is the economically strongest forces which must be challenged and tamed; their strength is such that their resistance is certain to be, in fact already is, powerful and tenacious. In England the battle has most probably been won. Under the stress of war all industries had to be controlled and their operation centrally planned in outline. Although they remained in 'private ownership', the actual meaning of that phrase became altered so that many rights which owners had previously enjoyed were taken over by the State. Soon after the war a Labour Government came to power before the owners had got back all their old freedom of action; and it now seems very unlikely that they will ever do so. The dense industrial system built up on this island, much too small to feed the working hands required, was always an unstable arrangement, and its former prosperity depended largely on a huge accumulation of overseas investments; now that we have spent these in fighting two wars, the adjustment of our industrial machine is going to remain for many years too tricky a matter to be trusted to the hit-or-miss operation of the law of supply and demand and the vagaries of private profit.

The same is true of the whole of Europe, including France. But the process has not gone so far in the United States, and no one can say whether the essential control over production will pass into the hands of the consumers peacably, or whether it will come to a fight. At present the capitalist interests seem to have been successful in persuading most Americans that socialism is the work of the devil, instead of merely meaning control by the people. In any case, Americans occupy such a huge chunk of the fertile temperate zone, in comparison with their numbers, that they are far removed from the practical need to make the best use of their resources which faces Britain and France. Their drive towards socialism will come (apart from war) from a breakdown in the profit machinery, such as produced the pre-war Depression.

It is impossible here to discuss all the detailed plans which have been put forward for the rationalisation of society, and in any case the details will have to be worked out in practice by trial and error. But there are some general results which seem to follow from the scientific view which are worth mentioning.

Perhaps the most important is that scientific investigation makes it quite clear that, just as we were finding it impossible to get into use much of the wealth which we created, so we were also wasting our human resources to almost the same degree. For instance, in a pre-war study of the English educational system,[55] it was found that, among the able pupils, 10% came from upper-class families (large business-men and professional men), about 40% from the middle-classes (shopkeepers, clerical workers and skilled workmen), and 50% from the lower-classes (manual workers). All these could have been greatly profited by a higher education; but although in the upper-class practically every able child, and an equal number of not-so-clever ones as well, obtained such an education, in the lower-classes less than a quarter of the bright ones went through a secondary school, and in the middle-classes the situation was only slightly better. Over half the children who would have profited from a good education failed to get one. This wastage of intellectual ability is an inefficiency comparable to the shockingly low standards of nutrition which we have put up with. No technically competent biologist would tolerate it in his rat colony for a week, and a pig business which treated the swine like that would be bankrupt in a couple of seasons, unless it was subsidised by some Pig Board or other.

It is clear that a scientific society, which had at heart the good of the community as a whole, would enormously increase the opportunities for education. Post-war Britain is, indeed, making huge efforts to build more schools and to send more children to them and to the Universities. It is, on a long term view, perhaps the most important of all our enterprises at the present time. Further, a scientific society will have to be organised so that the able, having received an adequate training, could come to the top and use their abilities in the positions of power and responsibility. This would mean an enormous change in the distribution of power; the children of our upper-classes, the professions and big-business-men, contribute only one-tenth of the community's supply of able men, but a much higher proportion of its influential people. If power were to be in the hands of those best able to use it sensibly, nine-tenths of it would go to the offspring of the middle and lower-classes. In a scientific society, 'power would be in the hands of the people'.

In practical terms the problem is more complicated than is suggested by that slogan. Although it is the great bulk of the population, the lower and middle-classes, which give rise to the greater part of our potentially able children, they suffer at present from the lack of proper education. It is not clear that it would be a good plan to put too much power directly into the hands of partially trained people, however clever. On the other hand, if it is left where it was, largely in the control of a richer and more privileged class, there is no reason why the wastage of ability should not continue more or less indefinitely.

One of the possible ways of getting over this difficulty has been followed in Russia. During the Revolution, in which power was taken over by the workers, the way in which the power was used was actually decided by a highly-trained group of men. The policy of the workers was vague; the Communists, acting on the workers' behalf, put it into definite and practical shape. It was lucky that such a group was available, and contained some men of genius – Lenin, Trotsky, Stalin and others. They at least did not believe in witches or the efficacy of crossing oneself as a way of dealing with difficult situations. They could, and did, build a society which provided opportunity for the able but poor man to become educated. They should soon be able to trust the common sense of the population sufficiently to throw the doors to power open to everyone, in a truly democratic way. In fact, they say they are already doing so, but news about Russia is so obviously prejudiced in one way or the other that it is impossible for the ordinary person to be certain what is happening.

The actual mechanism by which power will be distributed will depend too much on particular political circumstances for science to have any fixed ideas about it beforehand. But it is a necessary characteristic of a scientifically organised society that its class structure should be much looser than at present. This does not mean that classes will disappear entirely and everyone be on an exactly equal footing; but social esteem will be given to a man for different reasons than it is today; not because he has a stake in the country in the sense that he can put a good deal of its wealth into his own pocket, but because he seems likely to contribute more than most people to the aims which society has in view. And in any case the differences in social status between people will be much less permanent and unalterable when it is possible for everyone to develop his potentialities to the full.

'Race' is another of the ways of dividing up mankind which, very important in our civilisation at present, would be very much less so in a scientific world. Not that science believes that all men are racially alike. What we know about human heredity shows conclusively that they are not. But, firstly, the scientific races do not correspond with political groups; Frenchmen from Alsace are racially much more like Germans from Baden than like other Frenchmen from Normandy. None of the great nations consists of a pure race or anything approaching it, and there is therefore no justification for making a new nation for every little so-called 'racial' group, like the Slovenes or Ruthenians.

'What really binds people together,' says a scientific student of man, Ruth Benedict,[56] 'is their culture – the ideas and standards they have in common.' Where a group of people have a common culture which is somewhat different from that of their neighbours, they should, other things being equal, be able to live their own lives in the way they please; but that is no reason for allowing them to set up tariff walls around themselves, or to go to the absurd lengths of war. Modern means of communication and transport, bringing power from where it is cheap to places where raw materials are available, have enormously increased man's productive capacity. It is ludicrous that this should be stultified, and people go hungrier than they need, because different groups have different fashions in haircutting; but that was the condition of the world before the war. It would, perhaps, cause still more ribaldry among the brutal and licentious Gods of Olympus if the United States and the Soviet Union blow each other (and us) to bits with atomic bombs, not because either wants to, but because each is afraid the other is just going to; but that has a fair chance of being the condition of the world in this alleged peace.

With the defeat of the Nazis and their scientifically childish theories, nationalism rather than racism has become the most dangerous intellectual and moral drug which tempts the modern man to discard his precarious hold on sanity. Those capitalists who are fond of pointing at red bogeys – and those Communists who expose the aggressive machinations of the imperialists – should reflect that neither of their opponents would constitute any general danger provided they would only agree to stay at home. Economic theory cannot be a threat to regions of the world which do not accept it until its adherents try to spread it; and it can only carry the physical threat of war – which is what the world fears now – when it becomes allied with the central national governments which control the war-making machines. Those who wish to continue in their devotion to nationalism will, however, try to diminish the force of this by arguing that it is in the nature of capitalism (or Communism as the case may be) to take control of national power and use it as a weapon for ends which arise, not from nationalism, but from the economic theories, which are paramount. Nationalism, they urge, is an estimable sentiment – no danger to anyone – when it is not debased and brutalised by false social and economic theories.

This apology for nationalism is, in my judgment, neither wholly true nor wholly false. Nationalism is too complex a sentiment, and is involved with too many aspects of our emotional and political life, to be dismissed so simply. A scientific evaluation of it must start with a more serious attempt to understand its nature – and when a scientist speaks of 'understanding the nature' of something, he is thinking of a study of the causes which bring that thing about, and the effects which it produces. This book is no place to attempt a full scientific investigation of the modern concept of nationalism, but one can see almost at first glance that at least three different elements are usually lumped together under that name, each fundamentally important. There is, first of all, the sense of a national way of life – the feeling of being an Englishman, an American or a Frenchman, the sense of community for which one feels homesickness. That is something quite different from the conception of the nation as a unit of power, either as the sovereign ruling force in internal affairs, or as a piece on the chess-board of power politics. This power element is the second major factor in the concept of nationalism. The third which I should distinguish is one which is, perhaps, less generally recognised as distinct from the other two. It is the concept of the nation as an object of fierce patriotic devotion. It is the growth in recent times of the belief that the nation can properly demand the complete, or almost complete, submission of its individual members to the national purpose which has made nationalism into a world menace.

This element in the mixture is of tremendous importance even when it does not go so far as an outright admission that the individual exists for the nation, and not the nation for the individual. It was the root of the jingo imperialism of Britain forty years ago, in the time of songs like 'Land of Hope and Glory'; it is the root now of the paradoxical synthesis of isolationism and imperialism which has been achieved, in ways which are superficially different but fundamentally very alike, in America and Russia; it is the root also of much of the intransigence of political movements of liberation from Albania to Indo-China. It is, in fact, the most obviously important aspect of nationalism. We must be on our guard against thinking we have accounted for it, or even excused it, when we have in fact only discussed the other two factors; and this is a common mistake of discussions which treat of nationalism in general without analysing that concept into its constituent elements. It is essential to consider the specific role of this factor which we may call 'authoritarian nationalism', and to find out how, if at all, it is related to the other concepts which we commonly associate with it under the same name.

The study of 'national character', the first of our three elements, has recently been tackled with some success by anthropologists. They have investigated its nature in the truly scientific way – by analysing the causes which bring it about. The classical study in this field is that of Mead and Bateson on the Balinese[57] – a group of people chosen, not because of their intrinsic importance in the modern world, but because their national character, with which they feel at home, is one which seems exceedingly foreign, and therefore easy to grasp, to the English or American eye. The fundamental principle of Mead and Bateson's approach was to observe (with all the paraphernalia of cinema shots, dictaphone recordings and so on) the processes of social interaction by which the young Balinese child is moulded, by the adults and other children with whom it is in contact, into an example of the general pattern.

There is no room here to give anything like a complete picture of the Balinese, but it is worth while to pick out one or two points to serve as examples of the method. One of the first things that struck Mead and Bateson about the Balinese was their dependence on a very complex framework into which everything has to be fitted. They are continually referring to the points of the compass; the language in which they speak to one another varies according to the castes of the two speakers; and they use, simultaneously, weeks of one, two, three and so on up to ten days. 'In a strange village, where he does not know the cardinal points or the local customs, and if he does not know what day it is in at least three of the inter-cogging weeks, nor the caste and order of birth term (i.e. first, second or third child) for the person with whom he is trying to converse, the Balinese is completely disoriented. . . . If one takes a Balinese quickly, in a motor-car, away from his native village so that he loses his bearings, the result may be several hours of illness and a tendency to deep sleep.' This profound reliance on convention has many roots in the way Balinese children are brought up. One simple, but probably significant, fact is that a Balinese child is taught to walk by being given a framework which it walks round and uses as a support – if it wanders away it falls down. In contrast to this, we usually put our children in a square play-pen, which encourages them to keep to the outermost edges of their play-area instead of to the centre of it. The Balinese mother, in fact, gets very alarmed if her child strays out into an open space, and usually shouts some scare-word ('snake', 'wolf', 'dragon', or anything else which comes into her head) until the baby comes back in alarm. This method of controlling the child by frightening it is very characteristic of Balinese upbringing, though perhaps 'fright' is not quite the right word; 'startle' gives a rather better idea. The emotion is very common in grown-up Balinese: they have, quick, timorous reactions, rather like deer. And they obviously find this emotion not altogether disagreeable; after all, it was originally something they learnt by sharing it with their mother when they were in her arms.

It is easy to understand why the Balinese, so easily startled if they lose touch with their well-known background, find their greatest pleasure in gay and lively crowds. They like things to be, as they call it, '*rame*', a word which can be translated as 'noisily, crowdedly festive'. 'Pressed tightly against the steaming bodies of strangers, the air heavy with scent and garlic and spices, and many rare forms of dirt, sharing no single emotion with those so close to him, the Balinese watches the play and revels in the occasion, when he can stand completely remote in spirit, yet so close in body to a crowd.'

In that description of Balinese fun, the authors bring out another of the salient features of their national character – the lack of human warmth and personal contact. It is probably the most interesting thing about them, because it is in connection with the characteristic attitude to personal relations – whether they are seen on a basis of co-operativeness or of dominance and aggression, of rivalry or of emotional response – that the problem of national character is most pressing in the world today. And Bateson and Mead's analysis of this aspect of the Balinese character is particularly full and enlightening. They show how it is developed by a self-perpetuating system in which the adults have been brought up to be people who are bound, by their own characters, to bring children up just like themselves.

Balinese adults have been trained not to like warm, personal, human contact; they have learnt to withdraw, to sink back into themselves, if anything of the kind threatens. So they treat their children in just that way. Mothers play a great deal with their babies; in fact, according to our ideas they over-stimulate them and poke and tease them for the fun of making them smile and 'coo'. But as soon as the baby, or young child, begins to show any signs of real emotion, either of affection or rage, the Balinese mother turns to something else and leaves the baby's response hanging in the air. In the gay chattering family life, the child's emotions are continually being stimulated and then frustrated – until after a time the quite young toddler learns to keep them to himself.

There is a fairly definite sequence of stages in a Balinese baby's life. First he is the new-born of the family, his mother's most darling plaything; in fact, everybody is ready to tease him all the time. His mother will flirt with him in an openly sexual way, ruffling up his penis and saying how handsome he is; or, to tease him, she will borrow someone else's baby and pretend to have transferred her attentions to it. But it is all play; if the child reacts either by affection or rage, the mother turns away and starts laughing about something else. Then a new baby is born; and the previous favourite becomes still more a mere butt for teasing. By the time still another baby appears, is it any wonder that the little child has become a self-sufficient creature who never gives himself away? If the child is a girl, at this stage she begins to be used as a nurse to look after the newest baby; a boy soon takes charge of an ox, and spends his time herding in the fields where he has little contact with the rest of his family.

The process of damping down the child's spontaneous emotional reaction, of teaching it, above all, to avoid carrying anything to a climax, does not go entirely smoothly, as might be expected. Children who are just getting to the stage of complete control usually pass through a phase of extraordinary tantrums or of sulks. In a final attempt to get some reaction out of their mother, they go off into tempests of rage, yelling and throwing themselves about – or they sit and ostentatiously sulk. Neither tactic is often successful; the tantrum evokes no answering rage, and the sulking child is merely neglected. After a few years of this, the child has been formed into an individual who will treat the next generation of children just as it was treated itself – the system is self-supporting.

An analysis in similar terms could be made of other national characters; in fact the task was begun by American anthropologists during the war in order to guide the government services concerned with propaganda and morale (see for instance the article on Japanese character in Penguin *Science News I*). But these studies only help us to understand what one might call the 'community temperament' or 'climate of personality' characteristic of a human culture. They do not go far in accounting for the exclusive and aggressive features of nationalism at the present time. It is only in a few out-of-the-way cultures (such as some savage tribes) that the national character favours individuals who behave in the unprincipled and competitive manner exhibited by nations on the world stage. The seeds of the most dangerous aspects of nationalism must be looked for elsewhere than in those social processes which produce the simple feeling of being at home among one's neighbours.

The fact that nations are units of power, which we mentioned as the second major aspect of nationalism, has its roots, of course, in a long process of political development which has been often and copiously analysed in historical writings. As was stated earlier (Chapter VII), it seems plausible to suppose that, fundamentally, social power is derived from the production of goods and services of social value, and that the broad outline of political history, complicated though it is by the tortuous ways in which the human mind can cover up and distort the ultimate realities, should in the last analysis be expressed in terms of economic relationships. But nationalism, as we have seen, is by no means a matter of power politics alone; and although an economic analysis may reveal the sources of the political power which is now attached to nations, and even explain why it is to the national unit, and not to some other, that the power has been attached, it does not account for the actual geographical entities which nowadays constitute the nations. Before and between the last two wars, for instance, the fundamental economic structure of Western civilisation was based on groups, such as the capitalists, the industrial managers, the workers and so on, which spread through many different nations. But this internationality of economic structure had almost no effect on the power struggles of the nations. Even Britain, so often looked on as the centre of international imperialism, was quite unable to prevent other nations starting a war, although Britain's entire economic structure was founded on the essentially peace-like activity of world trade. Political power was in the hands of nations; and national boundaries did not coincide with boundaries of economic interest. Neither did they coincide simply with linguistic boundaries; Great Britain was one nation when it still included large numbers of Welsh and Gaelic-speaking people. Nor are national frontiers to be explained simply in terms of 'national character'; there is at least as much difference between a New Englander, a man from Chicago and a Texan as between a Flemish Belgian, a Dutchman and a German from Hamburg.

The demarcation of the earth's surface into regions, which

are now called nations, seems in fact to have been brought about largely by a series of historical accidents. A government, by some means or other, acquires power, happens to regulate the relations between people living in a certain area; that makes it the ruler of the internal affairs of its country. But this internal power does not necessarily mean that the national government is the seat of all power in external affairs affecting several nations. Through much of European history, for instance, the external power of national governments was no greater than, if as great as, that of the Church; and today in federations such as the United States or the Soviet Union, a great deal of internal power is in the hands of states which possess practically no external power at all. The pre-eminent position which the nation now occupies as the overwhelmingly most important seat of power in world affairs must be explained in some other way. And in fact the explanation must be sought in the third element in nationalism, which we called the authoritarian.

The origin of this third element is the imperfection of the human mechanism of social evolution. Man's society is founded on the training of the young in certain moral principles which embody the rules required to regulate the interaction between people in such a way that social life can continue. The forces of egoism which these principles have to restrain are very deep-seated; but in the human family and the relations between parents and children, man has evolved a mechanism which makes it possible to convey the essential moral principles at a very early age, when there is still a chance that they can gain control. The drawback of the mechanism is that it occurs mainly in the most early developed and primitive parts of the mind, which in adult life are usually hidden in the unconscious. Thus the strains and stresses which inevitably arise during the taming of the human animal into a well-behaved member of society often express themselves in highly devious ways. The most important of these maladjustments, from our present point of view, is a tendency to a morbid worship of some source of power which is substituted for that of our parents who originally imposed the social decencies on us. There is an almost universal craving to find some outside body or thing to which a man can attach his feelings of loyalty and submission. It may be a Church, or a less mundane God, it may be a political party, it may be an ideal (I shall argue in the last chapter that it might be science itself); but at the present day it is frequently a nation. The reasons for the fashionableness of the nation as a peg on which to hang these unconscious emotional needs require much more careful investigation than they have yet received. But the immediately important fact is that, among all the possible things which emotionally appear as sources of power, it is the nation which the spirit of our age chooses to pin its hopes and loyalties to.

It is this choice of the nation as the main emotional authority which poses the most difficult task for statesmanship today. Everyone admits the urgent necessity to set up an international authority with power to prevent the nations going to war. If this were a mere matter of constitution-making, there should be no great difficulty in devising a system which would guarantee the rights of all the peoples. We have the examples of Switzerland, the United States and the Soviet Union, in each of which men of the most diverse national characters live together under an authority which preserves a reasonable degree of order and at least prevents them going to war with one another. But any effective international authority of this kind must inevitably reduce the power of the present nations; and so long as men's unconscious emotions find satisfaction in the idea of national power, they will resist the growth of international government.

We still know far too little of the psychological and social mechanism involved to be able to offer a reliable prescription of how best to proceed. It seems fairly certain that nothing will be accomplished merely by stating (as Emery Reves has done so urgently in his book *The Anatomy of Peace*) that nations must give up some of their exclusive sovereignty. Possibly the historians, who have studied the transition from rule by the feudal nobility to rule by a single king, may be able to suggest how a new loyalty to a larger organisation can be brought into being. One suspects that the process will involve one of those self-reinforcing mechanisms so common in biological and social affairs; that men will not transfer their allegiance to a new authority unless that is already supported by considerable power, while the authority cannot obtain effective power until enough men have given it their loyalty.

A situation of that kind is not such a complete impasse as it suggests at first sight. But to move it, it is necessary not to rely on any single angle of approach, but to attack it from all sides simultaneously; organising and strengthening those sources of social power which stand behind it (in the present instance all the common people who hate war and the international economic and 'non-political' organisations) while at the same time helping forward the battles of the fledgling ruler (the United Nations) against its older competing sovereignties, and doing everything possible to weaken the hold which these out-of-date symbols (the nations) have over men's emotions.

Another major change in man's emotional orientation which seems to be demanded by the recent evolution of civilised society is the reduction in his taste for big-city life. There are many reasons why huge metropolitan agglomerations like London and New York seem to be unstable arrangements, doomed in the nearish future to break up and become scattered more widely over the land. They have been best discussed by Lewis Mumford in his book, *The Culture of Cities*. Here are some of them. When a city becomes too big, its land values rise so high that rents can only be earned if people are herded together in incredible congestion. Traffic in the rush hours is snarled up and jammed to such an extent that in New York, for instance, it is estimated that about £100,000 a day is lost on this account alone. The slums breed disease; health services in American cities of over a million cost three times as much per head as in smaller cities, but even then do not bring the mortality rates down so low as in them. The same is true of crime; large cities spend more on their police, and still have more criminals. And the city, which should provide the opportunity for wider and more varied human contacts, becomes, when it is a huge metropolis, a cold and unfriendly place, in which people can be more lonely than in a village.

Biologically, big cities are not self-supporting. Their birth-rate is too low. Metropolitan life does not encourage big families, and this applies to new industrial areas as well as to older cities. It is only the constant flow of people from the country into the towns which enables the big cities to remain constant in numbers, let alone to grow in the phenomenal way they have recently. But now the process is bringing about its own end.

The fashion for small families has spread from the town to the country, and the supply of possible new immigrants to the cities is tapering off. 'One is driven to ask,' write two students of the problem,[58] 'whether it is possible to maintain a stationary population in a community where urban congestion is sufficiently prevalent to dictate the *mores* of family life.' The cities are sucking the whole life out of the country and sterilising it. 'This metropolitan world,' says Mumford,[59] 'is a world where flesh and blood is less real than paper and ink and celluloid. It is a world where the great masses of people, unable to have direct contact with the more satisfying means of living, take life vicariously, as readers, spectators, passive observers; a world where people watch shadow-heroes and heroines in order to forget their own coldness or clumsiness in love, where they behold brutal men crushing out life in a strike riot, a wrestling ring or a military assault, while they lack the nerve even to resist the petty tyranny of their immediate boss.' That is pre-war stuff; how much better the people really are than this has been made magnificently obvious when they were given a chance to show their mettle in the defence of their homes against bombing. It is only the mad hustle of an overgrown city which, in the 'normal' times of peace, finds no use for the courage and endurance of air-raid wardens and fire fighters or the cheerful friendliness of the crowded shelters and tubes.

The reaction against city life has already started. It, too, can go in two ways, the Nazi or Fascist way, and the scientific. The Nazis wanted to take the unemployed out of the cities and turn them into peasants, to win a meagre living from the soil without the aid of farm machinery or supplies of power. The alternative way depends on the decentralisation of industry, making full use of modern means of distributing power wherever it is required, and of the best techniques of communication to keep widely separated factories or offices in touch with each other. The best example of it to date, perhaps, is the planned development of the valley of the Tennessee river in America under the Tennessee Valley Authority set up by the Government.

The huge tract of country drained by the Tennessee was in a bad way, constantly subjected to devastating floods, with its soil, unintelligently cultivated by poor and backward farmers, being washed and blown away by storms. The Authority,

T.V.A., tamed the floods by building dams, used the water-power to generate electricity, distributed this power in a grid to towns and villages, encouraging new industries to be set up in the local townships, and making it possible for the farmers to buy farm machinery and to cultivate their land more successfully and less wastefully. They explain their functions thus:[60] 'If a single private corporation could own the whole Tennessee watershed, it would naturally provide as many different services as could be produced from the river, collecting its charges for these services in various ways according to circumstances. Navigation privileges, water supplies, electricity would be sold directly to consumers; flood control and less tangible benefits would appear as increased rental values of real estate. In this way many services that could not be made profitable as separate business enterprises would be profitable if combined into a single trust. No such all-embracing private trust is likely to be allowed in our democratic system. The only all-embracing organisation is democracy itself. The people own the undivided rights in the Tennessee valley as well as in the rest of the nation. A job such as taming the Tennessee River, which is too big for a private corporation, must be managed by the agents of the people if it is to be done at all.'

Throughout the industrial world for the last fifty years, the units in which business and manufacture is organised have been growing larger and larger, until there are now many world-wide concerns with almost absolute powers within their field. It seems reasonably certain that this process must continue; it is dictated by technical advances, whose full benefit can only be reaped if they are used on a large and comprehensive scale. The economic organisation of the world is going totalitarian, and nothing can stop it, except perhaps a return to handicraft methods of production only capable of supporting a smaller population. What will happen to human freedom during this change? Must a realistic view of the world accept it as inevitable that as production becomes more organised men must also be regimented under central control, and thus lose their liberty?

Freedom is a very troublesome concept for the scientist to discuss, partly because he is not convinced that, in the last analysis, there is such a thing. Science works by discovering causes, and it finds difficulty in admitting that there is a free will, or desires and impulses which have no underlying cause. But leaving such philosophical questions on one side, science recognises that there is a feeling of freedom which men have, and that when they protest against the danger of losing their freedom they are not just talking hot air.

The 'freedom' men talk of really covers three or four different things. There is the freedom of not being restrained by law from doing certain things if one wants to; there is the freedom of being enabled to do one's best, to show initiative and to contribute something of one's own to society; and there is the freedom to be odd, to disagree with the majority and go one's own way.

The first freedom everyone recognises to be more or less of a farce at the present time. We are all free to dine at the most expensive hotels, to get the most highly-valued education society provides at the most fashionable public schools of England or private schools of America, to set up in business as a newspaper proprietor. Perfectly free to do so – if we have the wherewithal. The legal freedom to do such things is better than being prohibited from doing them, but it is not enough in itself. It has practical importance only if there is an opportunity to take advantage of it. As was said above, science is definite that ability is so widely spread in the community that it would be to the community's advantage to spread opportunity equally widely. Many more than half the able people in the country do not get a satisfactory training or scope for their talents, though there is no law against it. Their theoretical freedom will have to be made practical if this wastage is to be avoided.

The encouragement of initiative in an organised society is a more difficult proposition. I think nobody would question that it is desirable. But it is often argued that any Governmental or central control invariably leads to all the faults of bureaucracy – red tape and timid mediocrity. The question is usually discussed as though bureaucracy automatically went with public ownership, and efficiency with ownership by private business, but it is difficult to believe that the connection is really direct. It seems more likely that it is a matter of the size of the organisation; if it is too big for the efforts of an individual man to be noticed, there is no incentive for him to do his best. If the trend towards monopoly or totalitarian organisation continues, this will be the state of all production, equally whether in private or public hands. But there is no reason to suppose that the difficulty is insuperable. In Germany, for instance, some of the public enterprises such as the new autoroads showed at least as much boldness in conception and efficiency in execution as any private concern would have produced. And in America some of the big monopolies, such as the American Telephone and Telegraph, are as competent and up-to-date as anyone could wish. In fact, the most far-reaching and the most audacious of recent enterprises have, in actual fact, mostly been carried out not by private enterprise but by governments. For instance, the provision of stable civil government, and the largest irrigation system in the world, in the British India that was, or the development of the atomic bomb in the United States.

Probably there are two main factors to bear in mind when trying to combine initiative with large-scale organisation. If the community as a whole values and insists on efficiency, it will probably get it. Americans really believe whole-heartedly in having gadgets that work, and the A.T. & T., monopoly though it is, is a part of American society and shares the belief.

Perhaps it is a more important point that men seem not to put out their best unless they have competition. But the competition need not be for private profit, as it is today. Most people who can show initiative and drive enjoy doing so, and can be tempted to do so if success will bring them greater scope. They can compete for a wider field of action; in fact, even now this is probably a considerably more important psychological motive for efficiency in business than simple profit-making. The giants of industry, the tycoons as America calls them, have mostly long ago earned more than they can possibly spend; they keep on because they are interested, and it is a subsidiary fact that the only way they can keep on is to continue to make profits, which are not their primary motive. There is no reason why competition for scope should not be combined with large-scale organisation, even if competition for profit cannot be.

It might be claimed, however, that the freedom to compete with one's fellows, and to show initiative in doing so, is a comparatively superficial matter, appealing only to those tough types who worship success. The true freedom, to some people, is essentially a private and internal affair. It is the opportunity to express in work whatever creative urge one may possess. The essential thing for this type of freedom is that a man's daily activities should not only call on him continually to decide for himself what he should do to carry out his task, but should do this in a sphere exacting enough to bring out the best of which he is capable. This 'freedom for creativity' is much greater in some of the handicraft activities, such as farming, carpentry, even bricklaying, than it is in most jobs on a mechanised mass-production line. It is worth noting how completely it is distinct and separate from many of the other concepts also called 'freedom'; the workmen who sculptured the mediæval cathedrals had plenty of scope for their creative instincts, but very little to question the dogmas of the Church or to win 'success' by changing their economic status.

Freedom for creation is one of the deepest and most important social ideals. People who resent the limitations on it which have undoubtedly followed the industrial revolution often blame this result on the scientists, who invented the industrial techniques; forgetting that it was not the scientists, but the industrial magnates and the bureaucrats who organised how these techniques should be put into operation. Actually science is itself one of the most essentially creative activities; it is comparable in this to painting, writing and the other arts. The men who carry forward the work of science are as little likely as any group in the modern world to undervalue the freedom to create. In a world run according to scientific ideals, the most important objective would be to encourage all those faculties in which man has evolved furthest ahead of the other animals; and the capacity for creative achievement is the most outstanding of these.

The last sort of freedom, the freedom to be odd and unlike one's neighbour, is not, I think, a scientific value; at least I should not like to have to produce scientific reasons why a community which allowed it must be better for its members than one which did not. In the extraordinary variety of societies studied by anthropologists, some allow much more latitude to individuals than others, and it is not clear that the laxer ones are usually the more successful. In fact, Unwin[61] has recently argued the exact opposite – that too great freedom, particularly sexual freedom, lowers the energy and drive available for the important tasks of the community. I think he has hardly made out his case, although it is probably true that there is some limit beyond which it is not desirable to go in permitting people to do what they like regardless of what their neighbours think. But, at any rate in England and America, people value very highly their liberty to go their own way. Even if there is no particular scientific reason to support them, scientists are bound to be very sympathetic to such an ideal, because within their own technical field freedom to differ from the majority is one of the essential conditions of their work. Once destroy the freedom for any scientist to disagree with the majority, including his elders and betters, and scientific progress would come to a fairly immediate end. What scientists need and can give good reasons for in their own special field, they are not likely to want to deny to society as a whole in its more general affairs.

Science is often accused of being responsible for war, and that from two different angles. The less bitter accusation is that science, by providing the world with more devastating and efficient machinery, has increased the horror and beastliness of the wars which men have always engaged in. To a certain extent this is true; in so far as science increases man's power, it has enabled him to deliver heavier blows. But it is only fair to point out that it also strengthens his defence. War today extends over a wider area and involves more people than the wars of the past, but the armies suffered a very low proportion of casualties in comparison with the magnitude of the effects obtained. The most appalling slaughter was caused by the mass extermination of civilians, primarily in Eastern Europe, for which relatively primitive methods could be, and largely were, employed.

There is no denying that technical developments have brought war back again to civilians and women and children. These did not escape in uncivilised wars, such as the raids of Attila and his Huns, nor in more or less anarchic civil wars, such as those of the French and Russian Revolutions, but for most of the last few hundred years warfare has been in an intermediate stage of technical development which confined its effects to the actual combatant troops. Most people have therefore acquired the feeling that it is less brutalising to kill a man in uniform than a woman in ordinary clothes, although the more militant feminists point out, perhaps rightly, that this feeling is really a relic of the days when women were considered the weaker vessel, and valuable articles of property. But if it is actually a backward step to bring war home to the whole population instead of only to the soldiers, it must be admitted that it is the development of more scientific methods of war which has brought this about. And it is possible that modern war, even if less destructive of life, is more devastating in its effects on other human values than were the wars of the past.

The development of the atomic bomb, and still more of bacteriological warfare, will almost certainly make this situation look very different in the future. There can now, I think, be little doubt that these scientific 'advances' have outstripped defence methods so far as to make possible vastly greater casualties and far greater destruction of material resources than before; and if defence measures are attempted against them, these are likely to involve very great disturbance of normal civilised life. While it is clear that the general body of scientists must take a great deal of responsibility in this connection – and particularly, as I argue above on p. 20, for the positive action required to control the new weapons – yet it would not be fair that they should carry all the odium for having introduced these horrors. The scientists who produced the weapons did so under the orders of the authorities properly constituted by society. Their responsibility was, on the face of it, exactly similar to that of the bomber crews who executed the area bombing of German cities – a policy which, on military grounds, was open to grave and, in my not very well-informed opinion, probably decisive objection. The only grounds on which more blame could be given to the scientists than to the aircrew would be that scientists should be social leaders who cannot evade the duty of providing more definite guidance than is expected from ordinary man – and this is a thesis which is not generally accepted.

The accusation of increasing the horror of war is sometimes reinforced by a second, that science actually justifies war and endorses it as an essential part of human activity. Science is alleged to believe that all evolutionary advance proceeds through the struggle for survival, and to look upon war as the most decisive expression of this struggle among men. Both Nazis and Fascists made this claim, although the Nazis tended to justify war more as the final expression of man's service to the State than as a test of his individual worth. But Mussolini expressed the argument very plainly in his article on Fascism in the *Italian Encyclopædia*:[62] 'Above all, Fascism, in so far as it considers and observes the future and development of humanity quite apart from the political considerations of the moment, believes neither in the possibility nor in the utility of perpetual peace. . . . War alone brings up to their highest tension all human energies, and puts the stamp of nobility upon the peoples who have the courage to meet it.'

The argument is based on a complete misunderstanding of the scientific theory of evolution. The phrase 'struggle for existence' is a catchy slogan put over by Darwin, who was a genius at popularisation; but it is not to be taken literally. If a few thousand acorns start sprouting in the underbrush of a wood, there is a 'struggle for existence' between them, but they do not get up and hit one another over the head. The struggle is metaphorical, and the scientific doctrine of natural selection gives no direct sanction to human war as a method of evolution.

Leaving the general theory of evolution, and going into more detail about man's nature, it is obvious that he is not a vegetable, and does not, and should not, behave like an oak seedling. He seems to be naturally a pretty active creature, and, like other active mammals, only does well if given a chance to use his powers to the full. But the other animals for which this is true do not in general go to war, and do not evolve through its effects. Individuals may sometimes fight among themselves, but the main 'struggle' is against things of a different kind; to escape from other animals who prey on them, or to catch other animals to eat, or to deal more successfully with the inanimate world. There is no reason from natural history why man, simply because he is the sort of creature who must be doing something active, must band together in groups to kill other men. In fact, anthropologists know of tribes to whom the idea has never occurred. 'I myself,' says Ruth Benedict,[63] 'tried to talk of warfare to the Mission Indians of California, but it was impossible. Their misunderstanding of warfare was abysmal. They did not have the basis in their own culture upon which the idea could exist, and their attempts to reason it out reduced the great wars to which we are able to dedicate ourselves with moral fervour to the level of alley brawls. They did not happen to have a cultural pattern which distinguished between them.'

Far from considering war as the highest activity of man, or even as an inescapable trait in human nature, scientists, even more than most people in this country, find it a damnable, if unavoidable, interruption of their serious business. For science is concerned with man's real evolutionary advance, with giving him the mastery over things which will enable him to translate his wishes and aspirations into fact, and to overcome the obstacles which prevent the full development of his nature. This mastery can only be increased by fuller knowledge and understanding. The world is a safe which cannot be cracked by a sledge-hammer; one has to learn the codes which open the locks. It is only a hindrance, and not a way of advance, if the safe-breakers start fighting, or if, when the door is beginning to open, someone stands in the way to prevent the others going in.

These two points – that the 'struggle for existence' is a more subtle process than its name might suggest, and that man's behaviour cannot be thought of as a mere animal fighting – were clearly recognised by one of Darwin's most penetrating contemporaries. I like to picture him riding to hounds near Manchester, hat crammed well down, red beard blown back against his muddy pink, and thinking up some crack about the surplus theory of value; he was Frederick Engels, a bad 'un to beat. He pointed out that, as well as a selection due to the pressure of over-population, there is a selection by greater capacity of adaptation; and he drew attention to the fact that the crude idea of the struggle for existence was a reflection of current social theories into natural history and could not then be brought back again into sociology as an independent argument for competitive capitalism or for war. 'The whole Darwinian theory of the struggle for life is simply the transference from society to organic nature of Hobbes' theory of *bellum omnium contra omnes*, and of the bourgeois economic theory of competition, as well as the Malthusian theory of population. When once this feat has been accomplished . . . it is very easy to transfer these theories back again from natural history to the history of society, and altogether too naïve to maintain that

CHAPTER X

BELIEVING IN SCIENCE

There is one human characteristic which today can find a mode of expression in nationalism and war, and which, it may seem, would have to be completely denied in a scientific society. That is the tendency to find some dogma to which can be attached complete belief, forthright and unquestioning. That men do experience a need for certainty of such a kind can scarcely be doubted. As Voltaire pointed out, if God did not exist man would have to invent him. And there is certainly a difficulty for science here, since scientific belief is a quantitative affair; one believes in some things more, in others less, but in none with such absoluteness that evidence becomes irrelevant. How, then, can believing man be at home in a sceptical scientific world?

The problem arises on two scales, as it were. To take the minor one first: many people have the impression that scientific statements are always intolerably cautious, hedging and stuck-in-the-mud. It apparently seems impossible to them that anything of moment can be concealed within the heavy sentences of the conventional scientific vernacular. The opening phrases: 'While it is idle to speculate, the available evidence may be taken to indicate . . .' or 'The possibility must, it appears, be envisaged . . .' – serve like the slime of an oyster to slip them down the gullet unchewed and hardly tasted. But the answer to this is simply a jeering 'Think again'. The wary circumspection is called for precisely because the ideas which science advances are so outrageous. While the orthodox held, with confident certainty, a theory of dreams which must have been a bore already to the Pharaoh's courtiers in the time of Joseph, science hesitantly insinuated the devastating notion of the unconscious mind. It is passionately debated whether or not we can, after death, make tables go bump; and meanwhile the geneticists point out that anyone who is succeeded by an unbroken line of descendants is assured of the material immortality of some of his essential substance, his genes (apart from the uncontrollable process of mutation), but that the quantity is halved in each generation and there is no way of determining which half is preserved; and the tissue-culturist can promise, with his hand on his heart, but touching wood with one little finger, to provide actual immortality of the flesh (of some parts and only in bits). It is not the caution with which they are advanced which makes such scientific results seem unexciting; it is that they are too fantastically out of line with our cultural traditions to register.

Moreover, the nearer it gets to action, the more specific, though not necessarily the more unqualified, the advice of science becomes. When it comes to a practical matter of eating, Professor Haldane[65] consumes no more than three ounces of baking soda at a sitting, betting on the current physiological theory that that quantity would alter the acidity of his blood just less than the amount required to kill him; his colleagues could argue disinterestedly the likelihood of the theory proving inadequate. What science gives one is a specific and detailed instruction which has a probability of bringing about the desired result; and scientific action is normally and essentially based on probabilities rather than on certainties.

That would be true of its action in social matters also. The evidence that improved education and nutrition would raise the general intelligence of the lower-paid strata of society is not unassailable; but it is strong enough to make that view probable enough to be a basis for action, and science would not only unhesitatingly act on it, if it had the power, but must insist that a failure to do so is a gross betrayal of scientific ethics.

The odds at which an experiment becomes worth trying depend on how much harm it can do if it goes wrong, and especially on its liability to get out of control. It is this consideration which forces science to be unquestionably on the side of democracy. In a social experiment, which is performed in order to improve the lot of the members of the society, it is they, the experimental material, who must be the judges of whether the experiment is a success, and the ultimate agents who decide whether it should continue or not. Neither scientists nor any other particular group of people can decide it for them; the scientists' function is to suggest changes worth trying and to point out the benefits which the general populace may have overlooked.

But is it enough to be definite without being certain? The major objection to scientific scepticism is to its fundamental uncertainty. People may be willing to bet on a probability when things are going well, but when the world seems against them, and they are depressed and unhappy, they tend to turn for comfort to something which they feel they can believe in without reservations. This desire for an ultimate bedrock of stability – for something which, instead of being at the mercy of the mere chance happenings of the world, stands outside them, uninfluenced and constant – is a very important element in man's nature. It is often argued that it is so powerful that it is bound to frustrate any attempt to base a society on an empirical, sceptical attitude. It is pointed out that in Russia after the Revolution the peasants and common people, though on the whole not unwilling to give up their belief in the God of the Orthodox Church, did not remain free-thinking, but proceeded to canonise Lenin and pay their devotions to him. And one of the most fundamental troubles of Europe, as was pointed out earlier, is the lack of any secure system of beliefs on which people can rely. Is science, for all its logical consistency, in a position to satisfy this primary need of man?

Nowadays, mainly through the efforts of psycho-analysts, we know something about the psychological nature of the need to believe. Every human being starts life in complete dependence on its parents, and under their absolute control. The most important process in development is the movement away from reliance on the power and protection of the parents, and the acquisition of the capacity to deal with circumstances unaided. No one, at the present stage of human evolution, completes this movement. No one entirely escapes from the loving, commanding parent, who becomes built up into the structure of a man's mind under the disguise of a conscience or a God; who remains something which one can, perhaps, disobey, but whose power, such as it is, does not derive from this world, but is primary and unquestionable, standing outside the world and coming before it.

In the normal man this external authority, or super-ego, as it is called, is counterbalanced and checked by the rational part of the mind, which is developed to handle the outside world. There is little doubt that the direction of evolution is towards a restriction of the importance of the super-ego, and an increase in the power and freedom of reason. But the factors which determine the relative strengths of these two aspects of the human mind are not of a kind which one can expect to change very rapidly. One could not abolish the super-ego by teaching children a set of ideas different to those they learn at present. The only hope of doing so would be by a complete change in the whole nature of the relationship between parents and children, which would not be at all easy to bring about.

It seems, then, that we have to reckon with a tendency of mankind to make themselves a super-ego which they can, from time to time when it suits them, believe in with an unreasoning devotion. But the substitute father may take many different forms; if it cannot be God it may be Lenin. Its nature is by no means an unimportant matter, and is probably much more easy to influence than its existence. It may be jealous, mysterious and violently opposed to reason ; but it might, on the other hand, be wise, tolerant and more reasonable than our weak minds can imitate; it might be that very process of evolution whose direction determines science's ethical attitude. There need, in fact, be no irreconcilable conflict between the rational and irrational parts of man's mind; he can have an unreasonable belief in the value of reason.

It is therefore quite possible for man's irrational impulse to find a way of satifying itself even in a scientific society. The scoffer suggests that, however empirical and objective the society claims to be, one will discover that faith has put in an appearance in it. But the reply is, What of it? Why shouldn't it be a helpful faith?

A scientific society, however, would not be founded on faith. The super-ego would be the interloper. Most societies of the past have, I think, claimed the opposite, that they are based on a religious doctrine. It was an emotional belief in feudal

Christianity which held the mediæval world together; which laid down the rules which prevented the wars from being too devastating, kept economic power within bounds by the doctrine of the 'just price', and determined the structure of society in the hierarchy stretching from the King, God's Anointed, to the lowest villein. But a glance at the history books will convince anyone that the super-ego did not have it all its own way. Plenty of men laid their beliefs on one side and used their wits to look after their own interests. In fact, if they had not done so, civilisation would have remained static and fixed for ever in the shape it had when the beliefs were consolidated.

The mediæval world was one theoretically founded on irrational beliefs, and in practice considerably invaded by empirical reason. A scientific world would be the other way round; theoretically based on empiricism, in practice undoubtedly entertaining many unproved beliefs. Its great advantage would lie in the fact that it is to the application of reason to the external world that civilised man owes his evolution from the savage. In societies founded on emotional beliefs, reason has to fight bitterly for recognition, let alone for power to put its conclusions into practice. In a scientific world this same reason, the only method of human progress, would have become the officially recognised basis of the whole social scheme.

When one looks towards a scientific society, and considers how the super-ego can be accommodated and made use of within it, it is necessary to realise that we are at present in a very bad situation in this respect. Our great-grandfathers pictured their super-egos in terms of a God. They might, and frequently did, disobey his authority; he might be jealous, narrow-minded, cruel and generally disagreeable, although he might also be just and loving; but whatever his characteristics, a God has a certain scale. He is never simply mean, dingy and crotchety. And he dwells comfortably far away, and can be disregarded, when necessary, with comparative impunity.

Modern man seems in danger of losing both these advantages. The symbolic parent, whose authority we fear, is usually not called a God at all these days, and certainly does not deserve the title. The bogey that dominated the inter-war years was a much more bloodless creature – an old grey man muttering to his cronies in a club, a disapproving neighbour peeping through the curtain. It is called They – They won't like that, They won't approve, They don't do it like that, They say that's Bolshie. And They are essentially impersonal, negative, nondescript and incredibly suspicious. Their quiet, insinuating voice recommends firmly against action, except sometimes when They approve a violent expression of hatred. No thunderbolts issue from Their Olympus, only solicitors' letters warning against the consequences of pursuing the contemplated line of action; They are as unjovial as any gods could be.

That sort of a super-ego would be no good to a scientific society. It is not true, whatever you may have been led to believe, that the progress of science demands that everyone should always be right, and no society which takes caution as its main virtue will get anywhere, however scientific its thinking. But the pendulum seems to be already swinging away from Them, towards some new and livelier embodiment of authority.

It is, as They would be the first to point out, extremely hazardous to speculate on what kind of super-ego we shall make for ourselves in Their place. But even They might almost approve the conventional suggestion that we must be ready for the pendulum to swing to the opposite extreme. Are we in for a Romantic Revival? for one of those apparently paradoxical periods when the super-ego lends its authority to an ideal which emphasises spontaneity and naturalness? There is some evidence which can be interpreted in that way. It can be found in the cultural sphere. Painters, like John Piper, who were most conscious of the need to find some more solid 'tree in the field', are tending to go in that direction, to paint pictures in which the prime aim is romance, and the purely æsthetic advances of the abstract period are of secondary importance, incorporated where possible, but left to look after themselves. Poets like MacNiece defy Them:[66]

> No.
> You cannot argue with the eyes or voice;
> Argument will frustrate you till you die,
> But go your own way, give the voice the lie,
> Outstare the inhuman eyes. That is the way.
> Go back to where you came from and do not keep
> Crossing the road to escape them, do not avoid the ambush,
> Take sly detours, but ride the pass direct.

These are only hints, indications of a possible trend which opinion may take. My guess – and, I confess, my hope – is that they need to be taken seriously. For I do not think that an increase in the importance of a romantic view of life would make it any more difficult to achieve a scientific society. The substitution of a romantic ideal for our recent pallid and inhibited one would, I think, in itself simply release enormous potentialities of action which have been suppressed. It would be incompatible with the myriad vested interests, large and small, which sat on our heads like tin-pot or cast-iron lids. They would go flying.

The important question is what would emerge, and that would depend on what kind of romantic ideal we had adopted. As I pointed out before, almost any kind of ideal can be twisted to serve as the basis of a Fascist system. Not every kind is suitable for a scientific society; no anti-intellectual ideal would serve. But science has plenty of romance of its own to offer. In fact, its romance is so obvious and outstanding that it is traditionally the first enthusiasm of boys; to become an engine-driver used to be a child's first choice, nowadays perhaps to become a pilot. And the possibilities which science is opening up are so manifold and so large in scope that an enthusiasm of the same kind could provide the most important practical aims for a grown-up man; to take a responsible part in the carrying out of a great scientific enterprise, such as the development of the Tennessee valley, with its dams and power-houses and the social organisation which goes with them, would be no mean or uninspiring objective. Or if we wish to consider the matter in less immediate and concrete terms, I have already in an earlier chapter pointed out that science is not ethically neutral; it does imply a certain type of moral outlook. It has, in fact, something to say about the most important questions of the world, and it could therefore be a candidate for the position of super-ego.

The argument, like all good arguments in this finite world, has come back from the other side to the point it started from. I had suggested that one might have a scientific society, officially based on the practice of empirical reason; but I argued that the other side of man's nature would have to be satisfied by a belief in some authority, a thrill for some romance. We have

SUMMARY

The civilisation of the last few centuries is disintegrating; it suffered a stroke in 1914 and has been moribund ever since. It served its members indirectly, relying on the half-understood laws of private profit and supply and demand, and its whole system of beliefs and values was appropriate to such a mechanism. The type of economic system which will replace it is already clear in outline: it will be centralised, and totalitarian in the sense that all the major aspects of the economic development of large regions are consciously planned as an integrated whole. What is not clear is the character of the social ideals and values which will determine the purposes for which the system will be used.

The Nazi and Fascist totalitarian systems were fundamentally pointless; apart from the simple ideal of increasing the power already in the hands of a certain clique, they point to no ultimate direction for human striving. To the man left isolated in the world by the disappearance of the traditions he relied on, they offer only the gimcrack comfort of a rigmarole of emotive words; blood, soil, the Fascist spirituality. They would use the economic evolution towards an organised society to bring about a cultural regression to the Stone Age; and this is inevitable in any society in which a totalitarian economic system is used primarily for the benefit of a restricted group.

The application of conscious control to the economic functioning of society is actually a step forward along the path of man's evolutionary advance. Throughout the whole history of life on the earth, science observes an increase in the accuracy with which animals can perceive their surroundings, and a progressive development of the brain structures which enable them to reach a practical understanding of it. This general trend defines the direction which is forwards. It does more: it makes clear the method by which advance can be brought about – by action based on sensitive examination of the facts and rational inference from them. The mastery of this method is not child's play; it is the ideal of scientific culture.

The inter-war period has been a time of pupation; the capitalist world has been in the chrysalis of the Great Depression, breaking down its caterpillar economics and culture and preparing to emerge as a quite different-looking butterfly. The artists and writers who have been busy liquidating the outworn traditions have not formulated at all clearly the new culture which will succeed them. But the only part of their work which seems constructive and progressive is derived, unconsciously and tortuously it may be, from the scientific attitude. This is an attitude whose final standard of value is an observed process of evolutionary advance; it judges things not 'for themselves', but only for the effects which they produce on the rate of advance; no other considerations are relevant, however much traditional emotion may be attached to them. From this standpoint, the good of a society is measured by the good of its individual members; no special importance attaches to the welfare of particular groups, such as the present property owners, and none at all to that of abstract entities such as 'The State'.

The rational economic system, at whose birth pangs we are already assisting, can only be fully utilised if it is infused by a culture whose method of approach is also rational, intelligent and empirical. Prim Science has so far neglected to confess to the world that he has begotten such an offspring on the harlot Humanities; but the infant culture is beginning to peep already – in its bastard vigour lies the only hope for an heir worthy of the civilisations of the past.

LIST OF REFERENCES

1 *Technological Trends and National Policy*, U.S. Govt. Printing Office, 1937; p. 65

2 Margaret Mead, *Sex and Temperament*, Routledge, 1935; pp. 111 *seq.* and 279

3 Cf. R. M. Titmus, *Poverty and Population*, Macmillan, 1938; p. 86

4 See *Science and Ethics*, ed. by C. H. Waddington, Allen & Unwin, 1942

4a Pablo Picasso, from a conversation with Christian Zervos, published in English translation in *The Painter's Object*, edited by Myfanwy Evans, Gerald Howe, 1937 (one of the few books about painting written by painters)

5 T. S. Eliot, 'The Love-Song of J. Alfred Prufrock', from *Poems* 1905–1925, Faber & Gwyer, 1926

6 T. S. Eliot, 'Preludes I', from *Poems*

7 T. S. Eliot, *The Waste Land*

8 E. E. Cummings, 'One 2', from *Is. 5*, Boni and Liveright, New York, 1926

9 E. E. Cummings, 'Five 2', from *Is. 5*

10 T. S. Eliot, 'Animula'

11 John Piper, *The Painter's Object*; p. 69

12 Paul Nash, *The Painter's Object*; p. 108

13 Max Ernst, *The Painter's Object*; p. 79

14 Pablo Picasso, *The Painter's Object*; p. 81

15 Aldous Huxley, from *Leda*, Chatto & Windus, 1920

16 W. H. Auden, from *Another Time*, Faber & Faber, 1940; p. 70

17 William Empson, 'Invitation to Juno', from *Poems*, Chatto & Windus, 1935

18 William Empson, from *Poems*

19 Stephen Spender, 'The Indifferent One', from *The Still Centre*, Faber & Faber, 1939

20 Cecil Day Lewis, from *The Magnetic Mountain*, Hogarth Press, 1930; p. 18

21 Cecil Day Lewis, *The Magnetic Mountain*; p. 34

22 John Crowe Ransom, from *Two Gentlemen in Bonds*, Knopf, New York, 1927
23 Salvador Dali, from *La Conquête de l'Irrationnel*, Editions Surréalistes, Paris, 1935
24 T. S. Eliot, *East Coker*, Faber & Faber, 1940
25 W. H. Auden, from *Another Time*
26 T. S. Eliot, from *East Coker*
27 W. H. Auden, from *Another Time*
28 Marcel Breuer, in *Circle*, Faber & Faber, 1937; p. 193
29 J. L. Martin, in *Circle*; p. 215
30 Raymond B. Fosdick, *The Rockefeller Foundation – A Review for 1939*, New York
31 Robert S. Lynd, *Knowledge for What?* Princeton University Press, 1939; p. 8
32 A. V. Hill, *Nature*, 1933; p. 952
33 R. S. Lynd, *Knowledge for What?*; p. 6
34 See *The Middlesbrough Survey & Plan*, by Max Lock and others. Published Middlesbrough Corporation, 1947
35 A. V. Hill, *Cambridge Review*, Nov. 8th, 1940
36 Erich Kästner, 'Ein Mann Gibt Auskunft', from book of the same name, Deutsche Verlags Anstalt, Stüttgart, 1930
37 Rainer Maria Rilke, from *Requiem für Eine Freundin*
38 E. Krieck, quoted by Robert A. Brady, *The Spirit and Structure of German Fascism*, Gollancz, 1937; p. 60
39 B. Mussolini, *The Doctrine of Fascism*, publ. in *Enciclopedia Italiana*, 1932; English translation in *Social and Political Doctrines of Contemporary Europe*, by M. Oakeshott, Cambridge University Press, 1939; p. 164
40 Bernhard Rust, quoted by Brady; p. 51
41 J. B. S. Haldane, *The Marxist Philosophy and the Sciences*, Allen & Unwin, 1938; p. 16
42 e.g. J. B. S. Haldane, ibid.; p. 14
43 V. I. Lenin, *Materialism and Empirio-criticism*, quoted by Haldane in reference 41
44 A. N. Whitehead, *The Concept of Nature*, Cambridge University Press, 1920
45 Quoted by Haldane in reference 41
46 J. B. S. Haldane, *Heredity and Politics*, Allen & Unwin, 1938; p. 126
47 Robert S. Lynd, *Knowledge for What?*; p. 66
48 Ditto; p. 60
49 See G. Bateson, *Naven*, Cambridge University Press, 1936
50 From a preliminary report by Tom Harrison
51 L. F. Richardson, 'Generalised Foreign Politics', *British Journal of Psychology Monograph*, 1939
52 Kenneth Fearing, from 'C Stands for Civilisation' in *Dead Reckoning*, Random House, New York, 1938
53 Ruth Benedict, *Patterns of Culture*, Routledge, New York, 1933; p. 273

54 G. Bateson, '*Naven*', and '*Culture Contact and Schismogenesis*', from *Man*, 1935; p. 199
55 See the studies of Gray and Moshinsky, published in *Political Arithmetic*, edited by L. Hogben, Allen & Unwin, 1938
56 Ruth Benedict, *Patterns of Culture*; p. 16
57 Gregory Bateson and Margaret Mead, *Balinese Character*, New York Academy of Sciences, 1942
58 Charles and Moshinsky, in *Political Arithmetic*; p. 159
59 Lewis Mumford, *The Culture of Cities*, Harcourt Brace, New York, 1938; p. 258
60 T.V.A. Pamphlet, *To Keep the Water in the Rivers and the Soil on the Land*, U.S. Govt. Printing Office, 1938; p. 5
61 J. D. Unwin, *Hopousia, or the sexual and economic foundations of a new society*, Allen & Unwin, 1940
62 B. Mussolini, *The Doctrine of Fascism*; p. 170
63 Ruth Benedict, *Patterns of Culture*; p. 31
64 Frederick Engels, *The Dialectics of Nature*, translated by Clemens Dutt, Lawrence and Wishart, 1940; pp. 19, 208, 236
65 J. B. S. Haldane, see 'On Being one's own Rabbit', in *Possible Worlds*
66 Louis MacNiece from 'Eclogue in Iceland', in *The Earth Compels*, 1938; p. 38

NOTES

page 6

'The great slump' was a period of economic stagnation – or worse – in all industrial countries, except the U.S.S.R. It began with a crisis of confidence among financiers in the U.S.A. The slowing of industrial production that followed led to the prolonged unemployment of millions of workers, many of them skilled.

page 6

'England now is a worse country to live in than it was then.' The incidence of disease, the general standard of living, the variety of life and the improved opportunities for the children of the poor do not bear out this statement if it is supposed to apply to the 1960's.

page 7

Stanley Baldwin, later Earl Baldwin of Bewdley (1867–1947), was leader of the Conservative Party for many years, and Prime Minister 1935–37. He was criticised for publicly advocating support for the League of Nations (the predecessor of the United Nations) but failing to support it once he was in power.

King Edward VIII of Britain (b. 1894), now the Duke of Windsor, was obliged to abdicate soon after he came to the throne because he insisted on marrying an American woman, Wallis Warfield (Mrs Simpson), who had already been married twice, and divorced.

page 8

Fictional broadcasts of 'invasions from space' have caused panic in U.S.A. on more than one occasion.

page 11

'Totalitarianism' here implies a society highly organised and controlled by a central government, and with a planned economy. See also page 145.

page 15

The importance of recognising the 'plasticity' of human behaviour is underlined by some current writings on human 'aggression'. These imply that we have an 'instinct' or 'drive' which impels us to be destructive, and that we 'inherit' this from our animal ancestors. In fact, although almost any human being can be provoked to violence, there is no evidence that normal people are violent without provocation. Some people are more easily provoked than others. The important question is what conditions, in childhood or later, make some people peaceful and others combative. (Some men, such as soldiers, are of course trained to be violent.) Ideas of a fixed 'human nature' distract attention from the ways in which the development of the behaviour of individuals responds to the circumstances in which they are reared.

page 18

The journal *Nature* has now given up its annual analyses of modern art.

page 20

On 'the bomb' read *Brighter than a Thousand Suns,* by Robert Jungk (Penguin Books, 1964).

page 30

On cubism, and much else, read *The Story of Art* and *Art and Illusion,* both by E. H. Gombrich (Phaidon Press, 1950 and 1960 respectively).

page 35

The W.P.A. (Works Progress Administration) was set up under the 'New Deal' of President F. D. Roosevelt (1882–1945). At a time of slump and unemployment federal funds were used to give work to large numbers of people, including artists.

page 36

The statement about profiles and full-face views is questionable. The artists of ancient Egypt painted profiles with one eye shown as if it were seen full face.

page 47

The 'opening stages of the war' refers to the Spanish Civil War of 1936–39. This war began with a revolt by the army against the government of the Spanish Republic. The rebels were aided by the Fascist governments of Italy and of Germany (the Nazis), those loyal to the republic by Russians and volunteers (the International Brigade) from Britain, France, U.S.A. and other countries. The republic was defeated and replaced by the dictatorship of Francisco Franco.

page 79

Trofim Lysenko, a Russian plant breeder, attacked modern genetics ('Mendelism') and tried to replace it with a system based on the notion (sometimes called 'Lamarckism') that 'acquired' characters are inherited. Lysenko and his supporters are now discredited in the U.S.S.R. as well as elsewhere.

page 87

The formal definition of a Communist is a member of a Communist party, for example, of the Communist Party of Great Britain.

'U.N.O.' properly U.N. or United Nations.

page 110

Rudyard Kipling (1865–1936), poet and author, wrote two short stories (among his many others) in which the world appears to be run in an authoritarian manner by an organisation chiefly concerned with its transport. Kipling evidently favoured some such prospect. Aldous Huxley (1894–1963), novelist and essayist, wrote a novel, *Brave New World*, about a future in which all babies grow at first, not in a mother's uterus, but in 'bottles', where they are artificially fed and cared for. Huxley's book was designed as a *warning* against possible future trends.

QUESTIONS FOR DISCUSSION

1. In 1968 a report was published in Britain on a recent decline in numbers of schoolchildren who choose to study science at the Universities. The following reasons for the change have been suggested, among others: (i) a degree in social science or 'arts' subjects can be got with less effort; (ii) science is taught in an uninspiring way, and demands too much rote-learning; (iii) the contribution made by scientists to destructive weapons has given science disagreeable associations. Do you think any of these reasons are the real ones? Are they good reasons for avoiding science?

2. Should learning science start at the age of five? And should biology be taught first?

3. Should *all* educated people have a good grounding in the methods and conclusion of the natural sciences?

4. It is said that there are 'two cultures', that of literary people and that of scientists. Is this division avoidable, and if so should it be avoided?

5. Should scientists, invited to take part in work on methods of germ warfare, ruining crops and mass destruction, accept, on grounds of patriotism, or refuse, on grounds of loyalty to mankind?

6. Should efforts to put man on the moon or on Mars continue, or would it be better to concentrate on providing more food, preventing disease, controlling populations, and other immediately practical ends?